Springer Series on
SIGNALS AND COMMUNICATION TECHNOLOGY

T0181422

SIGNALS AND COMMUNICATION TECHNOLOGY

**The Variational Bayes Method
in Signal Processing**
V. Šmídl and A. Quinn
ISBN 3-540-28819-8

Topics in Acoustic Echo and Noise Control
Selected Methods for the Cancellation of
Acoustical Echoes, the Reduction of
Background Noise, and Speech Processing
E. Hänsler and G. Schmidt (Eds.)
ISBN 3-540-33212-x

**EM Modeling of Antennas and RF Components
for Wireless Communication Systems**
F. Gustrau, D. Manteuffel
ISBN 3-540-28614-4

Interactive Video
Algorithms and Technologies
R. I Hammoud (Ed.)
ISBN 3-540-33214-6

Continuous-Time Signals
Y. Shmaliy
ISBN 1-4020-4817-3

Voice and Speech Quality Perception
Assessment and Evaluation
U. Jekosch
ISBN 3-540-24095-0

Advanced Man-Machine Interaction
Fundamentals and Implementation
K.-F. Kraiss
ISBN 3-540-30618-8

**Orthogonal Frequency Division Multiplexing
for Wireless Communications**
Y. Li (Ed.)
ISBN 0-387-29095-8

**Circuits and Systems
Based on Delta Modulation**
Linear, Nonlinear and Mixed Mode Processing
D.G. Zrilic   ISBN 3-540-23751-8

Functional Structures in Networks
AMLn – A Language for Model Driven
Development of Telecom Systems
T. Muth   ISBN 3-540-22545-5

**RadioWave Propagation
for Telecommunication Applications**
H. Sizun   ISBN 3-540-40758-8

Electronic Noise and Interfering Signals
Principles and Applications
G. Vasilescu   ISBN 3-540-40741-3

DVB
The Family of International Standards
for Digital Video Broadcasting, 2nd ed.
U. Reimers   ISBN 3-540-43545-X

Digital Interactive TV and Metadata
Future Broadcast Multimedia
A. Lugmayr, S. Niiranen, and S. Kalli
ISBN 3-387-20843-7

Adaptive Antenna Arrays
Trends and Applications
S. Chandran (Ed.)   ISBN 3-540-20199-8

**Digital Signal Processing
with Field Programmable Gate Arrays**
U. Meyer-Baese   ISBN 3-540-21119-5

**Neuro-Fuzzy and Fuzzy Neural Applications
in Telecommunications**
P. Stavroulakis (Ed.)   ISBN 3-540-40759-6

SDMA for Multipath Wireless Channels
Limiting Characteristics
and Stochastic Models
I.P. Kovalyov   ISBN 3-540-40225-X

Digital Television
A Practical Guide for Engineers
W. Fischer   ISBN 3-540-01155-2

Speech Enhancement
J. Benesty (Ed.)
ISBN 3-540-24039-X

Multimedia Communication Technology
Representation, Transmission
and Identification of Multimedia Signals
J.R. Ohm   ISBN 3-540-01249-4

Information Measures
Information and its Description in Science
and Engineering
C. Arndt   ISBN 3-540-40855-X

Processing of SAR Data
Fundamentals, Signal Processing,
Interferometry
A. Hein   ISBN 3-540-05043-4

continued after index

Riad I. Hammoud (Ed.)

Interactive Video

Algorithms and Technologies

With 109 Figures and 8 Tables

 Springer

Dr. Riad I. Hammoud
Delphi Electronics and Safety
World Headquarters
M/C E110, P.O. Box 9005
Kokomo, Indiana 46904-9005
USA
e-mail: *riad.hammoud@delphi.com*

ISSN: 1860-4862

ISBN 978-3-642-06976-5   e-ISBN 978-3-540-33215-2

Springer is a part of Springer Science+Business Media.
springer.com

© Springer-Verlag Berlin Heidelberg 2006
Softcover reprint of the hardcover 1st edition 2006

Cover design: design & production, Heidelberg

To Ghada

Preface

My purpose in writing *Interactive Video: Algorithms and Technologies* is to respond to an increased demand for a complete scientific guide that covers concepts, benefits, algorithms and interfaces of the emerging interactive digital video technology. This technology promises an end of the days of dull videos, linear playback, manual video search, and non-consent video-content. Many futurists and I share the view that interactive video is becoming the future standard of the most attractive video formats which will offer users non-conventional interactive features, powerful knowledge-acquisition and teaching tools, efficient storage, as well as non-linear ways of navigation and searching.

In recent years, video content has skyrocketed as a result of decreasing cost of video acquisition and storage devices, increasing network bandwidth capacities, and improving compression techniques. The rapid expansion of Internet connectivity and increasing interest in video-on-demand, e-learning and other online multimedia rich applications, makes video as accessible as any static data type like text and graphic. The "open video" market that enables consumers to buy and rent digital videos over Internet, is taking another important boost as video distribution is becoming available for people on "the move". Now they can select, download and view various video types on their consumer electronics devices like mobileTV and video-playing iPods.

In light of these advances, video-content providers look at their daily demand for increased video storage as a valuable business assets and concern themeselves with how to easily manage, quickly access, efficiently present, browse and distribute videos. I believe this book will provide solutions to most of these problems by presenting a framework of automatic methodologies for transforming conventional videos to interactive video formats. As opposed to manually decomposing, generating summaries, and locating highlights or events in a long surveillance video, this book will show which algorithms could be implemented to automatically perform such tasks without tedious, painful or expensive human interventions.

At the first glance of this book, reviewers suggested to use in the title the term "media" instead of "video". While this suggestion is valid since video is a type of media, I believe that the term video is more appropriate here because most chapters are dealing with video images that in some cases are joined with synchronized audio and text captions like music videos, feature-length movies, sports and news videos. Also, the emphasis I put on the word video is due in part to the fact that digital video is now the most popular medium in entertainment, teaching and surveillance, and yet it is still the most challenging data type to manage, at least automatically and in large quatity. My intensive search on the web for a definition of interactive video was not very rewarding. It turned out that different people with different backgrounds have their own definitions based on specific applications. That's why I wrote a brief chapter that places the reader in the context of this book, and it defines the terms, components and steps of *raw interactive video, interactive video presentation* and *interactive video database*. This book tackles major topics of the first and third forms of interactive video, while only briefly mentioning the data integration and synchronization issues that users face in preparing *interactive video presentations*. Other specialized books on XML and media presentation could help readers in understanding these two issues more deeply.

While it appears that a large portion of this book is on algorithms with roots in computer vision, mathematics and statistics, I have strived in each chapter to include the motivations of the work from different angles and application areas of interactive video, present previous related work, show experimental results on various video types, and discuss interactivity aspects as well as users' experiences with the developed application-driven interactive video technology. Throughout this book all chapters followed this style of presentation, however each one treated a different research topic and engineering problem. This book is organized in four distinctive parts. The first part introduces the reader to interactive video and video summarization. Chapter two presents effective methodologies for automatic abstraction of a single video sequence, a set of video sequences, and a combined audio-video sequence. The second part presents a list of advanced algorithms and methodologies for automatic and semi-automatic analysis and editing of audio-video documents. Chapter five presents a novel automatic face recognition technology that allows the user to rapidly browse scenes by formulating queries based on the presence of specific actors in a DVD of multiple feature-length films. In particular, chapter 6 presents a new efficient methodology for analyzing and editing audio signals of a video file in the visual domain, that allows analyzing and editing audio in a "what you see is what you hear" style. The third part tackles a more challenging level of automatic video re-structuring, filtering of video stream by extracting of highlights, events, and meaningful semantic units. In particular, chapter 8 presents a detailed example of the Computational Media Aesthetics approach at work towards understanding the semantics of instructional media through automated analysis for e-learning content annotation. The last part is reserved for interactive video searching engines, non-linear browsing and quick video navigational systems.

This book couldn't be accomplished without the excellent scientific contributions made by a number of pioneering scientists and experts from around the globe who had shown their enthusiasm and support for this book. I am so thankful for their participation and the support of their institutions: Ognjen Arandjelović with *University of Cambridge*; C. G. van den Boogaart and R. Lienhart with *University of Augsburg*; Yunqiang Chen with *Siemens Corporate Research*; Chitra Dorai with *IBM Research Center*; Marc Gelgon with *INRIA*; Andreas Girgensohn, John Adcock, Matthew Cooper, and Lynn Wilcox with *FX Palo Alto Laboratory*; Benoit Huet and Bernard Merialdo with *Institut EURECOM*; Baoxin Li with *Arizona State University*; Yong Rui with *Microsoft Research*; M. Ibrahim Sezan with *SHARP Laboratories of America*; Frank Shipman with *Texas A&M University*; Alan F. Smeaton, Cathal Gurrin and Hyowon Lee with *Dublin City University*; and Andrew Zisserman with *University of Oxford*. Their expertise, contributions, feedbacks and reviewing added significant value to this unique book. I was so happy to see how much progress have been made in this area since I have launched a startup on Interactive Video Broadcasting back in year 2001, on the campus of Grenoble, France. I would like to thank my former Ph.D. advisor, Roger Mohr, believed in the bright future of this technological activity. At that time, interactive video appeared as a laboratory research and development tool, but today the sheer number of research groups, software companies and funding agencies, working and supporting this technology is good indication that this technology will soon be ready for mass commercial market. I owe an immense debt to my former sponsor and institution, Alcatel Alshthom Research and INRIA Rhone-Alpes for their support of three and half years of research and learning experience on interactive video. With the appearance of video-playing devices like iPod and mobileTV, and advanced prototypes of imagers like MPEG-7 cameras, I believe interactive video technology will soon be ported to these devices because it will help users on the move to select efficiently the movies before downloading them, manage the video data on limited storage devices as well as searching and absorbing video contents in an efficient manner.

I tried to cover in this book both algorithms and technologies of interactive videos, so that businesses in IT and data managements, scientistis and software engineers in video processing and computer vision, coaches and instructors that use video technology in teaching, and finally end-users will greatly benefit from it. Engineers will find enough details about the methods to implement and integrate interactive video in their software to be used on the platform of their choice. The detailed study of the state-of-the-art methods and the remaining issues to solve would benefit Master and Ph.D. students, as well as scientists in the field. Business mangers, video analyzers, sport coatches and teachers will learn innovative solutions to optimize time and human resources during the processing, editing and presenting of video assets.

Contents

Part I Introduction

Introduction to Interactive Video
Riad I. Hammoud 3
1 Introduction 3
2 What is an Interactive Video? 4
3 Video Transformation and Re-structuring Phases 6
   3.1 Video Structure Components 6
   3.2 Toward Automatic Video Re-structuring 9
      3.2.1 Shot Boundary Detection 9
      3.2.2 Key-frame Detection 10
      3.2.3 Object Detection 11
      3.2.4 Intra-Shot Object Tracking 13
      3.2.5 Recognition and Classification of Video Shots
           and Objects 14
      3.2.6 Video Events, Highlights and Scenes Detection 15
      3.2.7 Text Detection in Video Images 15
      3.2.8 Video Searching and Browsing Engines 16
   3.3 Establishing Navigational Links 17
4 Rhetoric of Interactive Video Playback 18
   4.1 Synchronization of Processed Content Playback 19
   4.2 Design of End-Users Interfaces 20
5 Examples of Authoring and Navigational Systems 20
6 Book Organization 24

Automatic Video Summarization
Benoit Huet, Bernard Merialdo 27
1 Introduction 27
2 State of the Art in Video Summarization 27
   2.1 Dynamic Summaries 28
   2.2 Static Summaries 29

2.3   Summary Evaluation 30
3   A Generic Summarization Approach 31
   3.1   Maximum Recollection Principle 31
   3.2   Illustration 32
   3.3   Maximum Recollection Principle Experiments 33
4   Multi-Video Summarization 34
   4.1   Definition 35
   4.2   Multi-Video Summarization Experiments 36
5   Joint Video and Text Summarization 37
   5.1   Joint Video and Text Summarization Experiments 38
6   Constrained Display Summaries 39
7   Home Video Network Interaction 39
8   Conclusion 41

Part II Algorithms I

Building Object-based Hyperlinks in Videos:
Theory and Experiments
Marc Gelgon, Riad I. Hammoud 45

1   Description of the Problem and Purpose of the Chapter 45
2   Particularities of Object Detection and Tracking Applied
    to Building Interactive Videos 47
3   Detection of Objects 48
   3.1   Introduction 48
   3.2   Motion-Based Detection 49
   3.3   Interactive Detection 49
4   Object Tracking 49
   4.1   Design and Matching of Object Observations 50
   4.2   Probabilistic Modelling of the Tracking Task 51
   4.3   Example of a Object Tracking Technique with Failure Detection 53
         4.3.1   Object Tracking by Robust Affine Motion Model
                 Estimation 53
         4.3.2   Automatic Failure Detection 55
5   Building Object-Based Hyperlinks in Video 57
   5.1   Appearance Changes in Video Shot 57
   5.2   Gaussian Mixture Framework for Classifying Inter-Shots Objects 59
         5.2.1   Object Classes Registration Interface 59
         5.2.2   Gaussian Mixture Modeling 60
         5.2.3   Classification of Non-Registered Object Occurrences 62
   5.3   Framework for Experimental Validation 63
6   Conclusion and Perspectives 64

Real Time Object Tracking in Video Sequences
Yunqiang Chen, Yong Rui 67
1   Introduction 67
2   Contour Tracking Using HMM 70
    2.1   Observation Likelihood of Multiple Cues 71
    2.2   Computing Transition Probabilities 72
    2.3   Best Contour Searching by Viterbi Algorithm 72
3   Improving Transition Probabilities 73
    3.1   Encoding Region Smoothness Constraint Using JPM 74
    3.2   Efficient Matching by Dynamic Programming 75
4   Combining HMM with Unscented Kalman Filter 76
    4.1   UKF for Nonlinear Dynamic Systems 77
    4.2   Online Learning of Observation Models 80
    4.3   Complete Tracking Algorithm 82
5   Experiments 83
    5.1   Multiple Cues vs. Single Cue 84
    5.2   JPM-HMM vs. Plain HMM 85
    5.3   UKF vs. Kalman Filter 86
6   Conclusion 87

On Film Character Retrieval in Feature-Length Films
Ognjen Arandjelović, Andrew Zisserman 89
1   Introduction 89
    1.1   Previous Work 91
2   Method Details 92
    2.1   Facial Feature Detection 92
        2.1.1   Training 93
        2.1.2   SVM-Based Feature Detector 94
    2.2   Registration 95
    2.3   Background Removal 96
    2.4   Compensating for Changes in Illumination 98
    2.5   Comparing Signature Images 99
        2.5.1   Improving Registration 99
        2.5.2   Distance 100
3   Evaluation and Results 101
    3.1   Evaluation Methodology 101
    3.2   Results and Discussion 101
4   Summary and Conclusions 103

**Visual Audio: An Interactive Tool for
Analyzing and Editing of Audio in the Spectrogram**
C.G. van den Boogaart, R. Lienhart 107
1   Introduction 107
2   From Audio to Visual Audio and Back Again 111
    2.1   Time-Frequency Transformation: Gabor Transformation 111

2.2   One Degree of Freedom: Resolution Zooming 115
2.3   Fast Computation 117
2.4   Visualization: Magnitude, Phase and Quantization 118
3   Manual Audio Brushing 119
3.1   Time-Frequency Resolution 119
3.2   Selection Masks 120
3.3   Interaction 123
4   Smart Audio Brushing 124
4.1   Audio Objects: Template Masks 125
4.2   Detecting Audio Objects 125
4.3   Deleting Audio Objects 127
5   Conclusion 129

Part III Algorithms II

Interactive Video via Automatic Event Detection

Baoxin Li, M. Ibrahim Sezan 133
1   Introduction 133
1.1   Related Work 134
2   Event-based Modeling and Analysis of Multimedia Streams - A
General Framework 135
3   Event Modeling via Probabilistic Graphical Models 136
4   Event Detection via Maintaining and Tracking Multiple Hypotheses
in A Probabilistic Graphical Model 139
5   Advantages of Using the Probabilistic Graphical Models 142
6   A Case Study: Football Coaching Video Analysis 144
6.1   American Football Coaching System 144
6.2   Deterministic Algorithm 145
6.3   Probabilistic Algorithm 147
6.4   Experimental results and performance comparison 149
7   A Novel Interface for Supporting Interactive Video 150
8   Summary and Future Work 153

Bridging the Semantic-Gap in E-Learning Media Management

Chitra Dorai 157
1   Introduction 157
1.1   Streaming Media Business Applications 158
1.2   Customer Relationship Management Applications 158
1.3   E-Learning Media Applications 159
2   Media Computing for Semantics 160
2.1   Computational Media Aesthetics 162
2.2   Understanding Semantics in Media 163
3   Computational Media Aesthetics at Work: An Example 164
3.1   Narrative Structures in Instructional Media 165

3.2 SVM-Based Instructional Audio Analysis 166
    3.2.1 Audio Feature Extraction 168
    3.2.2 Audio Classification Using Combinations of SVMs 169
    3.2.3 Experimental Results 169
3.3 Narrative Constructs in Instructional Media:
    Discussion Sections 170
    3.3.1 Audio Preprocessing 171
    3.3.2 Modeling Discussion Scenes......................... 171
    3.3.3 Discussion Scene State Machine 172
    3.3.4 Experimental Results 174
3.4 Visual Narrative Constructs in Instructional Media 175
    3.4.1 Homogeneous Video Segmentation.................... 176
    3.4.2 Detecting Picture-in-Picture Segments 177
    3.4.3 Detecting Instructor Segments 178
    3.4.4 Identifying Candidate Slide and Web Page Segments 178
    3.4.5 Discriminating Slide and Web-Page Segments 179
    3.4.6 Experimental Results 183
4 Conclusion 184

Part IV Interfaces

Interactive Searching and Browsing of Video Archives: Using Text and Using Image Matching

Alan F. Smeaton, Cathal Gurrin, Hyowon Lee 189

1 Introduction 189
2 Managing Video Archives 190
3 Físchlár-News: System Description........................... 194
4 Físchlár-TRECVid: System Description 198
    4.1 TRECVid: Benchmarking Video Information Retrieval 198
    4.2 Físchlár-TRECVid 200
5 Analysis of Video Searching and Browsing 204
6 Conclusions... 205

Locating Information in Video by Browsing and Searching

Andreas Girgensohn, Frank Shipman, John Adcock, Matthew Cooper, Lynn Wilcox 207

1 Introduction 207
    1.1 Detail-on-demand Video 207
    1.2 Video Search 208
2 Overview... 209
    2.1 Detail-on-demand Video 209
    2.2 Video Search 210
3 Hypervideo Summarizations 211
    3.1 Viewing Interface 213
    3.2 Determining the Number of Summary Levels 214

3.3   Segmenting Video into Takes and Clips 214
3.4   Selecting Clips for Summary Levels 215
3.5   Placing Links Between Summary Levels 216
4   Video Search User Interface..................................219
4.1   Query Result Visualizations: Story Collage, Shot Keyframe,
       Video Timeline 220
4.2   Feedback on Document Relevance: Tooltips and Magnified
       Keyframes 222
4.3   Overlay Cues: Visited Story, Current Playback Position,
       Query Relevance..223
5   Conclusions..224

References...225

Index 247

Part I

Introduction

Introduction to Interactive Video

Riad I. Hammoud

Delphi Electronics and Safety, IN, USA.
`riad.hammoud@delphi.com`

1 Introduction

In recent years, digital video has been widely employed as an effective media format not only for personal communications, but also for business-to-employee, business-to-business and business-to-consumers applications. It appears more attractive than other static data types like text and graphics, as its underlying rich content conveys easily and effectively the goals of the provider in means of image, motion, sound and text, all together presented to the consumer in a timely synchronized manner. Multimedia documents are more accessible than ever, due to a rapid expansion of Internet connectivity and increasing interest in online multimedia rich applications.

Video content has skyrocketed in light of fore mentioned advances, and as a result of increasing network bandwidth capacities, decreasing cost of video acquisition and storage devices, and improving compression techniques. Today, many content providers, like Google Video [118] and Yahoo [353], have created "open video" marketplaces that enable consumers to buy and rent a wide range of video content (video-on-demand) [292] from major television networks, professional sports leagues, cable programmers, independent producers and film makers. The evolving digital video archives include various types of video documents like feature-length movies, music videos from SONY BMG, news, "Charlie Rose" interviews, medical and e-learning videos, as well as prime-time and classic hits from CBS, sports and NBA games, historical content from ITN, and new titles being added everyday. Such video archives represent a valuable business asset and an important on-line video source to providers and consumers, respectively. The market is taking another important boost as video distribution is becoming available for people on "the move", where they can select, download and view various video types on their consumer electronics devices like mobileTV [227] and video-playing iPods [158]. In 2005, the number of hits for downloaded music data from iTunes on Apple's iPods reached 20.7 million [163]. This number is expected to increase with the appearance of video-playing iPods and mobileTVs.

Traditionally, video data is either annotated manually, or consumed by end-users in its original form. Manual processing is expensive, time consuming and often subjective. On the other hand, a video file provided in a standard format like MPEG-1 or MPEG-2 is played back using media players with limited conventional control interaction features like play, fast forward/backward, and pause. In order to remedy these challenges and address the issues of growing number and size of video archives and detail-on-demand videos, innovative video processing and video content management solutions – ranging from decomposing, indexing, browsing and filtering to automatic searching techniques as well as new forms of interaction with video content – are needed more than ever [324, 276, 130, 143, 303, 26, 278, 152]. Automatic processing will substantially drop the cost and reduce the errors of human operators, and it will introduce a new set of interactivity options that gives users advanced interaction and navigational possibilities with the video content. One can navigate through the interactive video content in a non-linear fashion, by downloading, browsing and viewing, for instance, only the "actions" of a film character that caught his/her attention in the beginning of a feature-length movie [194, 155, 130, 199, 299]. Retaining only the essential information of a video sequence, such as representative frames, events and highlights, improves the storage, bandwidth and viewing time.

Within the framework of the emerging "interactive video" technology, MPEG-4 and MPEG-7 standards [121, 124, 130, 293, 233, 275], this book will address the following two major issues that concerns both content-providers and consumers: (1) automatic re-structuring, indexing and cataloging video content, and (2) advanced interaction features for audio-video editing, playing, searching and navigation. In this chapter, we will briefly introduce the concept, rhetoric, algorithms and technologies of interactive videos, automatic video restructuring and content-based video retrieval systems.

2 What is an Interactive Video?

In order to simplify the understanding of interactive video environments, it is worth reviewing how people are used to viewing and interacting with video content. Current VCRs and video-players provide basic control options such as play/stop, fast forward/backward and slow motion picture streaming. The video is mostly viewed in a passive way as a non-stop medium where the user's interaction with the content is somewhat limited. For example, users cannot stop the video playback to jump to another place inside or outside the video document that provides related information about a specific item in the video like a commercial product, a film character, or a concealed object. Hence, the viewing of the video is performed in a linear fashion where the only way to discover what is next is to follow the narration and move through the video guided by seconds and minutes.

Such conventional techniques for video viewing and browsing seem to be inefficient for most users to get the crux of the video. Users ease and efficiency could be improved through:

1. providing a representative *visual summary* of the video document prior to downloading, storing or watching it. Alternatively, users could select the video based on just a few snapshots;
2. presenting a list of *visual entries*, like key-frames, hot spots, events, scenes and highlights, that serves as meaningful access points to desired video content as opposed to accessing the video from the beginning to the end; and,
3. showing a list of *navigational options* that allows users to follow internal and external links between related items in the same video or in other media documents like web pages.

Interactive video refers to nowadays uncommon forms of video documents that accept and respond to the input of a viewer beyond just conventional VCR interactive features like play and pause. For instance, in its basic form, interactive video allows users to pause the video, click on an object of interest in a video frame, and choose to jump from one temporal arbitrary frame to another where the selected object has appeared. Instead of being guided by seconds and minutes, the user of interactive video form navigates through the video in a very efficient non-linear fashion with options such as "next appearance", "previous scene" and "last event". Tentatively, the following definition of interactive video could be drawn:

Definition 1. *Interactive video is a digitally enriched form of the original raw video sequence, allowing viewers attractive and powerful interactivity forms and navigational possibilities.*

In order to ensure that humans do not perceive any discontinuity in the video stream, a frame rate of at least 25fps is required, that is, $90,000$ images for one hour of video content. This video content can be complex and rich in terms of objects, shots, scenes, events, key-frames, sounds, narration and motion. An original video, in MPEG-1 format, is transformed to interactive video form through a series of re-structuring phases of its content. Both the original video and the interactive video contain the same information with one major difference in the structure of the document. The original video has an implicit structure, while its interactive form has an explicit one. In an explicit structure, the hotspots and key-elements are emphasized and links between these elements are created. If such links are not established with items from outside the video content, then the produced document is called *raw interactive video*. For simplicity we will refer to it just as *interactive video*. Here we introduce two other extentions of interactive video documents: *interactive video presentation* and *interactive video database*.

Definition 2. *Interactive video presentation is a form of interactive video document that is centered on enriched video but is not exclusively video.*

In this context, the interactive video document contains not only the raw enriched video but also includes several kind of data in a time synchronized fashion. This additional data is added for two reasons: (1) enhancing the video content, and (2) making the video presentation self-contained. The type of data is determined by the author of the interactive video. This author would intelligently integrate the technique of the video producer or filmmaker with the technique and wisdom of the skillful teacher. The objective is to maximize the chances of convening the purpose and key information in the video well. Additional data can be documents of all types i.e. html, tables, music and/or still frames and sequence of images, that are available locally or remotely through the Web. Such types of interactive videos are also known as hyperfilm, hypervideo or hypermedia [151, 47].

Definition 3. *Interactive video database is a collection of interactive video documents and interactive video presentations.*

The main interactive video document is seen here as a master document which contains in itself a large number of independent interactive videos. Users can access the video database in two forms: searching and browsing. For instance, a sketch may be used to define an image in a frame. The system can look for that image and retrieve the frames that contain it. The user can also move the sketch within a certain pattern indicating to the system to find in a large interactive video database all sets of frames that represent a similar moving pattern for that object of interest (hotspot). As technologies such as object detection, tracking and image recognition evolve, it will be possible to provide better ways to specify hotspots in interactive video, which will be covered in more detail in the following sections. When browsing is not convenient for a user to locate specific information, he or she would utilize searching routines instead. The searching process would result in a small set of interactive video sub-documents that could be browsed and easily navigated.

3 Video Transformation and Re-structuring Phases

The transformation of an original video format to its interactive form aims at extracting the key-elements or components of the video structure first, and then creating links between elements of various levels of this structure.

3.1 Video Structure Components

The presentation of a document often follows a domain-specific model. Readers of a textbook expect to see the book content organized into parts, chapters, sections, paragraphs and indexes. Such structure is presented to the reader up-front in a table-of-contents. Unfortunately the structure of video documents is not explicitly apparent to the viewer. The re-structuring process aims at automatically constructing such a table-of-contents [276].

A video collection can be divided into multiple categories by grouping documents with similar structures together. Thus, feature-length movies, news, sports, TV shows, and surveillance videos have different structures, but with

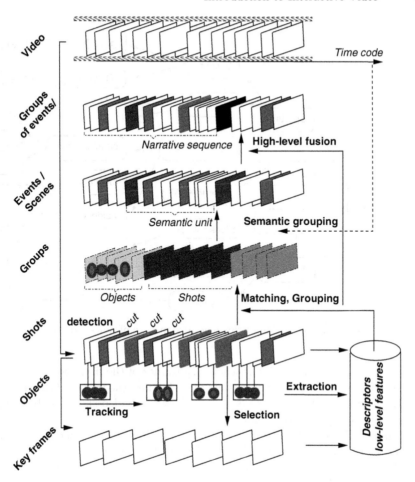

Fig. 1. Illustration of the structure of a feature-length movie

many basic elements in common. Figures 1 and 2 illustrate the general structure of movies and news videos.

The list of following components covers video structures of most types of video documents:

1. **Camera shots:** A camera shot is an unbroken sequence of frames recorded from a single camera, during a short period of time, and thus it contains little changes in background and scene content. A video sequence is therefore a concatenation of camera shots. A cut is where the last frame in one shot is followed by the first frame in the next shot.
2. **Gradual transitions:** Three other types of shot boundaries may be found in video documents: **fades**, **dissolves**, and **wipes**. A fade is where the frames of the shot gradually change from or to black. A dissolve is where the frames of the first shot are gradually morphed into the frames

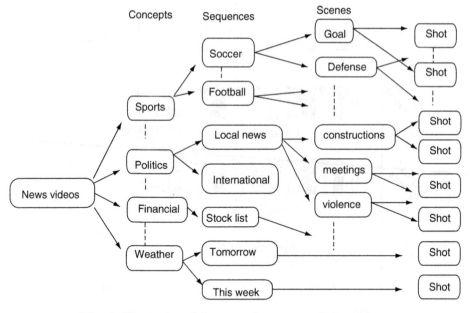

Fig. 2. Illustration of the general structure of news videos

of the second. And a wipe is where the frames of the first shot are moved gradually in a horizontal or vertical direction into the frames of the second.

3. **Key-frames:** A key frame is a still image of a video shot that best represents the content of a shot.

4. **Visual objects, zones:** Objects of interest in video documents are similar to key-words in text and html documents. They could be detected either manually or automatically in key-frames, or within individual frames of a video shot.

5. **Audio objects:** An audio object could take several shapes ranging from a clip of music to a single word.

6. **Text objects:** A text object is defined as the joint text to the video sequence, such as, footnotes and superimposed text on images.

7. **Events:** An event is the basic segment of time during which an important action occurs in the video.

8. **Scenes:** A scene is the minimum set of sequential shots that conveys certain meaning in terms of narration.

9. **Cluster of objects:** A cluster is a collection of objects with similar characteristics like appearance, shape, color, texture, sounds, etc.

10. **Narrative sequence:** A narrative sequence projects a large concept by combining multiple scenes together.

11. **Summary:** A video summary of an input video is seen as a new brief document which may consist of an arrangement of video shots and scenes, or an arrangement of still key-frames.

#1   #30   #90   #91

Fig. 3. A cut between two camera shots is observed at frames 90-91

These above components are labeled either as *low-level components* or *high-level components*. The first category includes video shots, audio-visual objects, key-frames and words, while the second contains events, scenes, summaries, and narrative sequences.

3.2 Toward Automatic Video Re-structuring

Manual labeling and re-structuring of video content is an extremely time consuming, cost intensive and error prone endeavor that often results in incomplete and inconsistent annotations. Annotating one hour of video may require more than ten hours of human effort for basic decomposition of the video into shots, key-frames and zones.

In order to overcome these challenges, intensive research work has been done in computer vision and video processing to automate the process of video re-structuring. Some interactive video authoring systems [127, 157] offer tools that automate shot detection, zone localization, object tracking and scene recognition.

3.2.1 Shot Boundary Detection

Several methods have been proposed for shot boundary detection in both compressed and non-compressed domain [38, 36, 13, 37, 134, 347, 356, 52]. Pairwise comparison [368] checks each pixel in one frame with the corresponding pixel in the next frame. In this approach, the gray-scale values of the pixels at the corresponding locations in two successive frames are subtracted and the absolute value is used as a measure of dissimilarity between the pixel values. If this value exceeds a certain threshold, then the pixel gray scale is said to have changed. The percentage of the pixels that have changed is the measure of dissimilarity between the frames. This approach is computationally simple but sensitive to digitalization noise, illumination changes and object motion. As a means to compensate for this, the Likelihood ratio, histogram comparison, Model-based comparison, and edge-based approaches have been proposed. The Likelihood ratio approach [368] compares blocks of pixel regions. The color histogram method [37] compares the intensity or color histograms between adjacent frames. Model-based comparison [134] uses the video production system as a template. Edge detection segmentation [364] looks for entering and exiting edge pixels. Color blocks in a compressed MPEG stream [34] are processed to find shots. DCT-based shot boundary detection [16] uses differences in motion encoded in an MPEG stream to find shots. Smeaton et al. [297, 246, 38] combine color histogram-based technique with edge-based and MPEG macroblocks methods to improve the performance of each individual method for shot detection. Recently, Chen et al. [54] proposed a multi-filtering technique that combines histogram comparison, pixel comparison and object tracking techniques for shot detection. They perform object tracking to help to determine the actual shot boundaries when both pixel comparison and histogram comparison techniques failed. As reported in [54], experiments on a large amount of video sequences (over 1000 testing shots), show a very promising shot detection precision of greater than ninety-two percent and recall beyond ninety-eight percent. With such a solid performance, very little manual effort is needed to correct the false positives and to recover the missing positives during shot detection. For most applications, color histogram appears to be the simplest and most computationally inexpensive technique to obtain satisfactory results.

3.2.2 Key-frame Detection

Automatic extraction of key-frames ought to be content based so that key-frames maintain the important content of the video while removing all redundancy. Ideally, high-level primitives of video, such as objects and events, should be used. However, because such components are not always easy to identify, current methods tend to rely on low-level image features and other readily available information instead. In that regard, the work done in this area [369, 377, 18, 341] could be grouped into three categories: static, dynamic and content-based approaches. The first category selects frames at a fixed time-code sampling rate such that each nth frame is retained as a key-frame [237]. The second category relies on motion analysis to eliminate redundant frames. The optical flow vector is first estimated on each frame, then analyzed using a metric as a function of time to select key-frames at the local minima of motion [341]. Besides the complexity of computation of a dense optical flow vector, the underlying assumption of local minima may not work if constant variations are observed. The third category analyzes the variation of the content in terms of color, texture and motion features [369]. Avrithis et al. [18] select key-frames at local minima and local maxima of the magnitude of the second derivative of the composed feature curve of all frames of a given shot. A composed feature is made up of a linear combination of both color and motion vectors. More sophisticated pattern classification techniques, like Gaussian Mixture Models (GMM), have been used to group similar frames [377], and close appearances of segmented objects [132], into clusters. Depending on the compactness of each obtained cluster, one or more images could be extracted as key-frames. Figure 5 shows extracted key-appearances from the video shot of Fig. 4 using

Fig. 4. Sample images of a tracked vehicle in a video shot of 66 frames

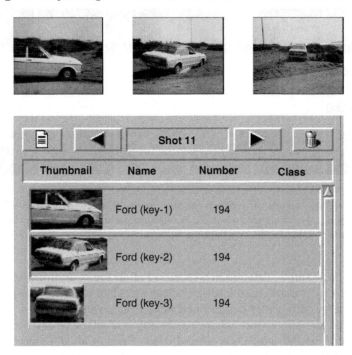

Fig. 5. Extracted three representative key-frames (top) and corresponding key-appearances (bottom) of the presented video sequence in Fig. 4, using the method of [132]

the method of [132]. In this example, the vehicle is being tracked, and modeled with GMM in the RGB color histogram space.

3.2.3 Object Detection

Extracting the objects from the video is the first step in object-based analysis. Video objects or hot spots are classified into static zones and moving objects. The purpose of object detection is to initialize the tracking by identifying the object boundaries in the first frame of the shot. The accuracy of automatic object detection depends on the prior knowledge about the object to be detected as well as the complexity of the background and quality of video

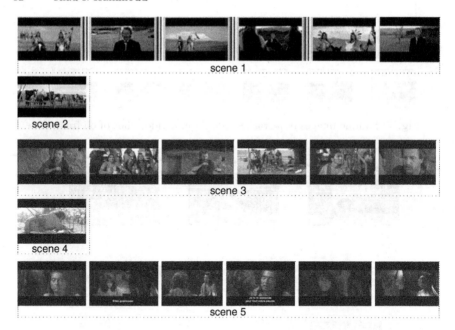

Fig. 6. Semantic scenes of a clip from the movie "Dances with Wolves": "first meeting" (scene 1); "horses" (scene 2); "coffee preparation" (scene 3); "working" (scene 4); and "dialog in the tent" (scene 5).

signal. As an example, detecting the position of an eye in an infrared driver's video [133, 136] is easier than locating pedestrians in a night-time sequence of images. The eye has a very distinct and unique shape and appearance among people while the appearance, shape, scale and size characteristics may vary widely among pedestrians. On a related topic, Prati et al. [256] recently proposed an object detection method that differentiates between a moving car and its moving shadows on a highway.

Several efforts were undertaken to find objects belonging to specific classes like faces, figs and vehicles [202, 240, 254, 120, 333]. Recent systems like the Informedia Project [155] include in their generation of video summaries the detection of common objects like text and faces. Systems such as *Netra-V* [79] and *VideoQ* [50] used spatio-temporal segmentation to extract regions which are supposed to correspond to video objects. Others [184, 269, 191] have resorted to user-assisted segmentation. Since it is easier to define an object by its boundary box, accurately delineating an object is generally advantageous in reducing clutter for the subsequent tracking and matching phases. In [269], the color distributions of both the object and its background are modeled by a Gaussian mixture. Interestingly, this process is interactive, i.e. the user may iteratively assist the scheme in determining the boundary, starting from sub-optimal solutions, if needed. Recent work goes further in this direction

[191], by learning the shape of an object category, in order to introduce it as an elaborate input into the energy function involved in interactive object detection.

In the interactive video framework, it is crucial that the tracking algorithm can be initialized easily through a semi-automatic or automatic process. Automatic detection methods may be categorized into three broad classes: model-based, motion-based (optical flow, frame differencing), and background subtraction approaches. For static cameras, background subtraction is probably the most popular. Model-based detection approaches usually employ a supervised learning process of the object model using representative training patches of the object in terms of appearances, scales, and illumination conditions, as in a multi-view face detector [202] or a hand posture recognition algorithm [344]. In contrast, a motion-based detection consists of segmenting the scene into zones of independent motions using optical flow and frame differencing techniques [240]. In many applications, the knowledge-based, motion-based and model-based techniques are combined together to ensure robust object detection. Despite all of these efforts, accurate automatic object detection is successful only in some domain specific applications.

3.2.4 Intra-Shot Object Tracking

Tracking is easier and faster than object detection since the object state (position, scale, boundaries) is known in previous frame, and the search procedure in target frame is local (search window) rather than global (entire frame). The tracking task is easier when tracked objects have low variability in the feature space, and thus exhibit smooth and rigid motion. In reality, video objects are often very difficult to track due to changes in pose, scale, appearance, shape and lighting conditions. Therefore, a robust and accurate tracker requires a self-adaptation strategy to these changes in order to maintain tracking in difficult situations, an exist strategy that terminates the tracking process if a drift or miss-tracking occurs, and a recovery process that allows object reacquisition when tracking is lost.

In literature, three categories of methods could be identified: Contour-based, Color-based and Motion-based techniques. For tracking of non-rigid and deformable objects, geodesic active contours and active contour models have been proved to be powerful tools [247, 181, 320, 250]. These approaches use an energy minimization procedure to obtain the best contour, where the total energy consists of an internal energy term for contour smoothness and an external energy term for edge likelihood. Many color based tracking methods [66, 27, 166, 249, 378] require a typical object color histogram. Color histogram is a global feature used frequently in tracking and recognition due to its quasi-invariance to geometric and photometric variations (using the hue-saturation subspace), as well as its low computation requirements. In [27], a side view of the human head is used to train a typical color model of both skin color and hair color to track the human head with out-of-plane rotation. The third category of tracking methods tends to formulate a tracking problem as the estimation of a 2D inter-frame motion field over the region of interest [110].

Recently, the concern in tracking is shifting toward real-time performance. This is essential in an interactive video authoring system where the user could initialize the tracker and desire quick tracking results in the remaining frames of a shot. Real-time object tracking is addressed in more detail in Chapter 4.

3.2.5 Recognition and Classification of Video Shots and Objects

Once the shots, key-frames, and objects are identified in a video sequence, the next step is to identify spatial-temporal similarities between these entities in order to construct high-level components of the video structure like groups of objects [131] and clusters of shots [275]. In order to retrieve similar objects to an object-query, a recognition algorithm that entails feature extraction, distance computation and a decision rule, is required. Recognition and classification techniques often employ low-level features like color, texture and edge histograms [24] as well as high-level features like object bobs [45].

An effective method for recognizing similar objects in video shots is to specify generic object models and find objects that conform to these models [89]. Different models were used in methods for recognizing objects that are common in video. These include methods for finding text [222, 375], faces [271, 286] and vehicles [286]. Hammoud et al. [131, 128] proposed a successful framework for supervised and non-supervised classification of inter-shot objects using Maximum A Posteriori (MAP) rule and Hierarchical classification techniques, respectively. Figure 8 illustrates some obtained results. Each tracked object is being modeled by a Gaussian Mixture Model (GMM) that captures the intra-shot changes in appearance in the RGB color feature space. In the supervised approach, the user chooses the object classes interactively, and the classification algorithm classifies remaining objects into the object class according to the MAP. In contrast, the unsupervised approach consists of computing the Kullback distance [190, 106] between Gaussian Mixture Models of all identified objects in the video, and then feeding the Hierarchical clustering algorithm [173] the constructed proximity matrix [169, 264]. These approaches seem to substantially improve the recognition rate of video objects between shots. Recently, Fan et al. [96] proposed a successful hierarchical semantics-sensitive video classifier to shorten the semantic gap between the low-level visual features and the high-level semantic concepts. The hierarchical structure of the semantics-sensitive video classifier is derived from the domain dependent concept hierarchy of video contents in the database. Relevance analysis is used to shorten the semantic gap by selecting the discriminating visual features. Part two of this book will address in detail these issues of object detection, editing, tracking, recognition, and classification.

3.2.6 Video Events, Highlights and Scenes Detection

In recent years, research towards the automatic detection and recognition of highlights and events in specific applications such as sports video, where events and highlights are relatively well-defined based on the domain knowledge of the underlying news and sport video data, has gained a lot of attention [182, 54, 199, 201, 198, 272, 281, 361]. A method for detecting news reporting was presented in [213]. Seitz and Dyer [289] proposed an affine view-invariant trajectory matching method to analyze cyclic motion. In [310], Stauffer and Grimson classify activities based on the aspect ratio of the tracked objects. Haering et al. [122] detect hunting activities in wildlife video. In [54], a new multimedia data mining framework has been proposed for the detection of soccer goal shots by using combined multimodal (audio/visual) features and classification rules. The output results can be used for annotation and indexing of the high-level structures of soccer videos. This framework exploits the rich semantic information contained in visual and audio features for soccer video data, and incorporates the data mining process for effective detection of soccer goal events.

Regarding automatic detection of semantic units like scenes in feature-length movies, several graph-based clustering techniques have been proposed in the literature [357, 275, 125]. In [126, 127], a general framework of two phases, clustering of shots, and margin of overlapped clusters, has been proposed to extract the video scenes of a feature-length movie. Two shots are matched on the basis of color-metric histograms, color-metric auto-correlograms [147] and the number of similar objects localized in their generated key-frames. Using the hierarchical classification technique, a partition of shots is identified for each feature separately. From all these partitions a single partition is deduced based on a union distance between various clusters. The obtained clusters are then linked together using the temporal relations of Allen [7] in order to construct a temporal graph of clusters from which the scenes are extracted [126]. Figure 6 depicts the results of the macro-segmentation process adopted in [126, 127] to extract semantic units from a feature-length movie.

In spite of such extensive existing work, there still has been a lack of progress in the development of a unifying methodology or framework for temporal event or scene detection, and in the development of a general metric for measuring performance. Part three of this book will address these issues in details.

3.2.7 Text Detection in Video Images

Text is a very common object in video. It can be treated as an object in itself or as a feature mark of an object (e.g. name on a football jersey, or a Taxi sign). In broadcast video, caption text is overlaid on scenes during video editing. The uniform editing of text for specific portions of TV programs (e.g. text to introduce people interviewed on CNN) make some text objects visually similar.

Many methods have been introduced to detect and locate text segments in video frames [210, 42, 366, 345, 363, 209, 305]. Two categories could be identified: component-based methods and texture-based methods. Within the first category, text regions are detected by analyzing the geometrical arrangement of edges or homogeneous color/gray-scale components that belong to characters [363]. Smith detected text as horizontal rectangular structures of clustered sharp edges [305]. Using the combined features of color and size range, Lienhart identified text as connected components that have corresponding matching components in consecutive video frames [209]. The component-based methods can locate the text quickly but have difficulties when the text is embedded in a complex background or touches other graphical objects [345]. Within the texture-based category of methods, Jain [363, 167, 168] has used various textures in text to separate text, graphics and halftone image regions in scanned gray-scale document images. Zhong [363, 374] further utilized the texture characteristics of text lines to extract text in gray-scale images with complex backgrounds. Zhong located candidate caption text regions directly in DCT compressed domain using the intensity variation information encoded in the DCT domain [363]. Those texture-based methods decrease dependency on the text size, but they have difficulty in finding accurate boundaries of text areas. The two categories of methods described are limited to many special characters embedded in text of video frames, such as text size and the contrast between text and background in video images. To detect the text efficiently, those methods usually define a lot of rules that are largely dependent of the content of video. Because the video background is complex and moving/changing, traditional ways that try to describe the contrast between text and video backgrounds have difficulty detecting text efficiently. Therefore, it might be useful to synthesize both the traditional method using many locating rules and that based on statistical models for detecting and locating text in video frames.

3.2.8 Video Searching and Browsing Engines

Given the ability to automatically extract low-level and high-level components of the video structure, it is possible to search for video content at a number of different levels, including features, models, and semantics [327, 141]. The shot boundary description defines the basic unit of matching and retrieval as a video shot or segment.

Users may issue queries to the system using feature-based and text-based approaches such as query-by-example and query-by-keywords, respectively [51]. In query-by-example, the user selects an example key-frame, object-instance images, and video segments. With query-by-keywords, the user issues a text query. An image-plus-text search facility and a relevance feedback process for query refinement has been proposed in [69].

The query-by-example approach suffers from at least two problems. The first one is that not all database users have example video clips on hand. Even if the video database system interface can provide some templates of video clips, there is still a gap between the various requirements of different users and the limited templates that can be provided by the database interface. The second problem is that users may prefer to query the video database via the high-level semantic visual concepts or hierarchical summary by browsing through the concept hierarchy of video contents. The major difficulty for the existing video retrieval systems is that they are unable to let users query video via the high-level semantic visual concepts and enable concept-oriented hierarchical video database browsing [302].

Query-by-keywords is also used in some content-based video retrieval systems based on manual text annotation [104], [150]. The keywords, which are used for describing and indexing the videos in the database, are subjectively added by database constructionist without a well-defined structure. Since the keywords used for video indexing are subjective, naive users may not be able to find exactly what they want because they may not be so lucky to use the same keyword as the database constructionist did. Moreover, manual text annotation is too expensive for large-scale video collections.

Two chapters of this book will present latest developments in this area.

3.3 Establishing Navigational Links

After decomposing the video structure and making it explicit, the next step is to establish navigational links among its key-elements. These links would provide end-users powerful navigation possibilities into the video content such as play next "event" in a sport video, or "previous appearance" of a film-character in movie.

Traditionally, the notion of links is used in hypertext as a powerful way to associate related information. Today, this concept is widely employed in the interactive video and hypervideo framework. An interactive video document offers the viewer two types of links: "internal links" and "external links". In general, a link is established between a source node and a destination node. The term "node" is used to refer to an element or component of the video structure. For instance, a node could be a frame at particular moment, a detected face in a key-frame, a video scene of a specific length, etc.

Definition 4 (Internal-links). *Internal-links are established only between extracted nodes of the decomposed video. As an example, an internal link is established between two separate events in which the same person of interest appeared.*

Internal-links are the direct result of the extraction process of high-level elements of the video structure. The author of the interactive video may also edit these generated links and/or add new ones. These links may be internal or extremal.

Definition 5 (External-links). *External-links are established between identified nodes of the central video and nodes from outside the processed video. For instance, if the user clicks on a vehicle in the video, the external link would trigger a destination video clip that shows a 3D model of the vehicle and further information about its manufacturer, safety ranking, and other power features.*

The integration of external links makes the video content richer and thus more attractive to a viewer for grasping as much knowledge as possible about the projected story or embedded concept behind the played video. Nodes of external links may encapsulate any type of media documents such as text files, still frames or music clips. External links are integrated only during the preparation of interactive video presentations or hypervideos. The user could select a specific event and watch video clips of its news coverage: current, historical, foreign, partisan. The news clips would contain opportunities for action that appear at related moments. A separate video clip appearing next to the main news clip would offer a related clip such as another point of view, a recent news account of the same story, or a commentary on the rhetoric of news reports in the 1960s.

By definition, interactive videos contain only internal-links while interactive video presentations may contain both internal and external links. Both types could be seen as link-opportunities as they imply a window of time or space when an association may be active. Such link opportunities are video and video structure-level dependent. They appear dynamically and provide navigational pathways to related scenes, inside or outside the video document. However, they do not exist in text or graphic types of media since they are static. The message of text is present at all times on the screen, while the video's message depends on the changes in the images that occur over time. When reading a text file, the user sets the tempo of information acquisition. Then, while viewing a video document, the tempo is set by the machinery that makes the illusion of motion pictures and hence link opportunities appear in a brief window of time (say, three to five seconds) to pursue a different narrative path. If the user makes a selection, the narrative jumps to another place in or outside the video.

4 Rhetoric of Interactive Video Playback

In this section we will review important design steps of interactive video playback interface: synchronization of content presentation and user interaction modalities. The goal is to represent the processed video content to the user in a very intuitive and effective way.

4.1 Synchronization of Processed Content Playback

The issue of synchronization comes at the time of compiling all links and associated nodes together. It is the last step that the author of the interactive video form has to go through in order to define how the processed content would be presented to the end-user. As the end-user selects, for instance, a link opportunity that triggers a destination video clip, the synchronization process will determine whether to stop playing the original video until the destination video ends playing, or just play both videos simultaneously. If the source video pauses then it is important to determine what will happen when the destination video ends. It could start from the same point in time where it was paused or it could restart from the beginning. Also, when a user clicks on a key-frame it could be possible that the main video re-position itself to play the corresponding video shot. When a user desires to navigate the interactive video at the level of scenes, jumping to "next scene" may imply that the next scene clip replaces the source video. The shape and flow of interactive video presentation depends on the design of the synchronization process. With all these possibilities and the fact that different paths through the interactive video are possible, it is important that the narrative sequence and the whole presentation remain coherent. Such coherence could be managed through different techniques being adopted by different interactive video authoring systems.

The World Wide Web Consortium (W3C) group proposed recently the Synchronized Multimedia Integration Language (SMIL) to address the problem of synchronization of multimedia presentations. SMIL is completely complementary to the HTML language. While this latter is designed to describe the presentation of a Web page, SMIL language describes instead the presentation of the multimedia data that fills the Web page. SMIL defines the mechanism to compose multimedia presentations, synchronizing where and when the different media are to be presented to the end user over the Internet. SMIL is text-based code that tells the video player where to get and how to play the media elements in the movie. It is designed to support the layout of any data type and file format, or container file formats. It provides a way of taking different media and placing them relative to one another on a time line, and relative to one another on the screen. Since, it is an open language, it is capable of modifications and refinements.

Due to the complexity added by the inclusion of time and the rhetoric imposed by time-based media in an interactive video framework, SMIL and other available ISO languages like HyTime or XMT-O (Extensible MPEG-4 Textual Format), seem to have limitations in representing all interaction and presentation aspects of interactive video content. Thus a lot of effort is placed on the development of interactive video authoring systems.

4.2 Design of End-Users Interfaces

Interactive video requires a careful design of the end-user interface in order to efficiently present the enriched content of the video document to the end-user. The user needs to be made aware of extracted key-elements of the video structure, and the links that are established. The interface could be designed in different ways: providing selection bottom for each extracted level of the video ("events", "group of objects", etc), flashing wireframes within the video, changes in the cursor, and/or possible playback of an audio-only preview of the destination video when the cursor moves over the link space. Several large and overlapping wireframes could detract from the aesthetic of the film-like hypervideo content, yet the use of cursor changes alone requires the user to continuously navigate around the video space to find links. For example, opportunities within the scene could be represented by creating a visual grid around the characters or by changing the color properties to emphasize specific regions. Link opportunities within a shot could have different grid resolutions to provide a visual distinction for simultaneous spatio-temporal opportunities. Several interaction techniques, such as changes in the cursor shape and volume adjustment may depict different link opportunities. Changes in the cursor prompt the user to move around the video space [283]. As the engine plays a video sequence, the user may watch the narrative or skip to other parts of the video.

5 Examples of Authoring and Navigational Systems

Now we will present briefly some recent authoring systems and interactive video playback interfaces.

VideoPrep. A system called VideoPrep (see Fig. 7) was developed by IN-RIA and Alcatel Alshtom Research (AAR) between 1998 and 2001. It allows the author of interactive video to automatically segment the video into shots, key-frames, groups of objects (see Fig. 8) and video scenes [131, 128, 126, 127]. It also provides friendly user interfaces to edit the results of the automatic process, adding more objects, and creating internal and external links. The system provides some interactive tools to correct the results of the automatic clustering techniques: (1) *select/browse* clusters at different levels of the hierarchy, (2) *Drag* a badly classified object and *Drop* it into the appropriate cluster.

VideoClic. The produced interactive video is played-back with an end-user interface called "VideoClic" (see Figs. 9 and 10). This interface provides control options that allow users to navigate in a non-linear fashion by jumping from one scene to another, and from one object occurrence to another. The end-user clicks an object of interest (actor, car, ...), jumps to its next or previous occurrence in the video, plays the corresponding shot or the corresponding action, and discovers a related external link (webapge, video, music clip, etc.) The presented structure allows the end user to access the interactive video document as a book with a table-of-content. Here a non-linear

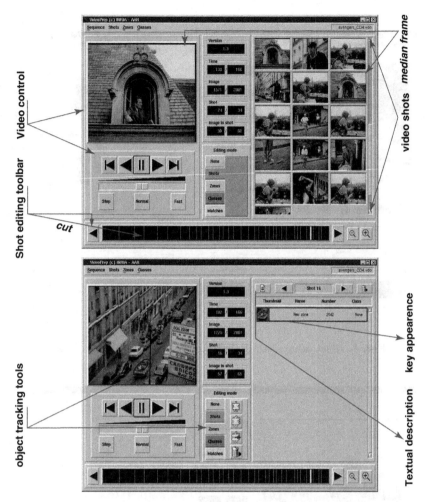

Fig. 7. Illustration of some re-structuring phases of VideoPrep system: *shot detection* (top) and *object tracking* (bottom)

navigation in the structure of the video is performed. The highlighted areas in the horizontal scroll bar of Fig. 9 indicates the shots where the object of interest has occurrences.

iVideoToGo. Intervideo company has released a tool that edits and filters the video content automatically [157]. They provided an intelligent detection technology that extracts the highlights from long, dull videos and turns them into a compact movie with fancy effects. Their new iVideoToGo product converts DVDs, Video Files to iPod Movies. It also supports the MPEG4 (object-based representation) and H.264 video formats.

Movideo SDK. Movideo is a tool from Arts Video Interactive [229] that allows creating hypervideo links on QuickTime, AVI and MPEG files. It also

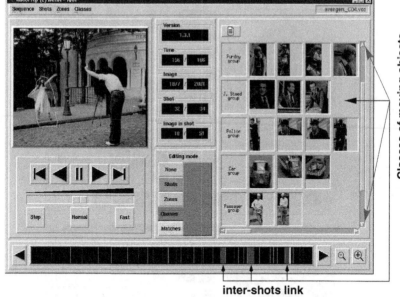

Fig. 8. Illustration of the *clustering* phase of VideoPrep system

Fig. 9. Snapshot of VideoClic interface

Fig. 10. Snapshot of VideoClic Java interface for interactive video playback

helps track either manually or semi-automatic trajectories (movements and transformations) of the actors and objects in a digital video sequence. Movideo SDK also can create hypervideos that may be integrated within other multimedia titles and presentations as ordinary videos. The system works with Macromedia Director, Asymetrix Multimedia Toolbook, mFactory mTropolis, Microsoft Visual Basic and programming languages (C, C++, Pascal, and Delphi). In 1997, over 37 million Shockwave players (Flash and Director combined) were successfully downloaded from Macromedia's web site. While Macromedia Director does not provide hypervideo functionality directly, other tools like Movideo add the hypervideo functionality to it.

VeonStudioTM technologies. Veon Inc. has deployed its VeonStudioTM system for authoring interactive video applications [92]. These tools are targeted for applications such as video shopping, interactive entertainment, training and distance learning. Veon adopted SMIL as the underlying multimedia format. Its V-Active HyperVideo Authoring Engine provides functionality to specify a hotspot in one frame and then recognize the image in subsequent frames, reducing the load of specifying the hot spot for every frame.

Throughout the chapters of this book several interactive video authoring and browsing systems will be presented in detail.

6 Book Organization

This book is organized in four distinctive parts. The first part introduces the reader to interactive video and video summarization. The next chapter will provide effective methodologies for automatic summarization and abstraction of a single video sequence, a set of video sequences, and a combined audio-video sequence.

The second part presents a list of advanced algorithms and methodologies for automatic and semi-automatic analysis and editing of audio-video documents. Chapters 3, 4 and 5 describe adequate solutions to the problems of automatic, robust and real-time object and face detection, tracking, recognition and classification. The last chapter of this part, presents an efficient methodology for analyzing and editing audio signals of a video file in the visual domain. The presented approach allows analyzing and editing audio in a "what you see is what you hear" style.

The third part addresses a more challenging level of automatic video restructuring, filtering of video stream by extracting of highlights, events, and meaningful semantic units. Chapter 7 proposes an automatic approach for event detection in the framework of interactive video. Video can be nonlinearly browsed event by event, or events can be used in generating semantic indexes for search and retrieval of video. Chapter 8 describes Computational Media Aesthetics as an approach to bridging the semantic gap and outline its foundations in media production principles. It presents a detailed example of the Computational Media Aesthetics approach at work towards understanding the semantics of instructional media through automated analysis for e-learning content annotation. This chapter describes how algorithms for identification and description of high-level narrative segments in instructional media are guided by production knowledge. I will show results from analysis of numerous educational and training videos from various organizations, thus demonstrating automated techniques for effective e-learning content management.

The last part is reserved for interactive video searching engines, non-linear browsing and quick video navigational systems. Chapter 9 and chapter 10 describe efficient and friendly users tools for browsing and searching through video to enable users to quickly locate video passages of interest. In many cases, it is not sufficient to search for relevant videos, but rather to identify relevant clips, typically less than one minute in length, within the videos. Several approaches will be offered for finding information in videos. The first approach provides an automatically generated interactive multi-level summary in the form of an interactive video presentation. When viewing a sequence of short video clips, the user can obtain more detail on the clip being watched. For situations where browsing is impractical, we present a video search system with a flexible user interface that incorporates dynamic visualizations of the underlying multimedia objects. The system employs automatic story segmentation, and displays the results of text and image-based queries in ranked sets of story summaries. Both approaches help users to quickly drill down to potentially relevant video clips and to determine the relevance by visually inspecting the material. Both of these are based on an analysis of the captured video for the purpose of automatically structuring into shots or higher level semantic units like TV news stories. Some also include analysis of the video for the automatic detection of features such as the presence or absence of faces.

Automatic Video Summarization

Benoit Huet and Bernard Merialdo

Institut EURECOM, 2229 route des crêtes, 06904 Sophia Antipolis, France.
benoit.huet@eurecom.fr and bernard.merialdo@eurecom.fr

1 Introduction

Due to the ever increasing number of multimedia documents one is potentially confronted with everyday, tools are eagerly awaited to ease the navigation through massive quantities of digital media files. Summaries provide an interesting solution to this problem. Indeed, by looking at a summarized version of documents one is able to quickly identify interesting or relevant documents.

In this chapter, we present a brief review of recent approaches to video summarization, and then we propose our approach based on the Maximum Recollection Principle. We show that this approach is supported by reasonable assumptions, and that this principle can be applied in diverse situations. In particular, we describe how it can be applied to the summarization of a single video sequence, a set of video sequences, and a combined audio-video sequence. For all these cases, we present some experimentation and discuss implementation issues for the corresponding algorithms.

2 State of the Art in Video Summarization

The automatic creation of multimedia summaries is a rather powerful tool which allows us to synthesize the entire content of a document while preserving the most important or most representative parts. Here, we concentrate on video summarization. In this respect, the creation of a video summary will result in a new document which may consist of an arrangement of video sequences or an arrangement of images. In other words, a video summary may take the form of a dynamic or a static document. The original document represented in such an abstract manner may be perused in various ways. It may help, for example, to get a quick feel about the content of a document or even about the general content of an entire database of multimedia documents. Another example of possible usage, particularly well suited for multi-episode TV series, is the ability to identify documents which have already been watched. Along the same line a video summary should enable the viewer to decide whether the content of the original is relevant or not.

This leads us to the obvious fact that a document can be summarized in a number different ways. Each of these individual summaries may have equal quality with respect to their intentional usage despite being different. This clearly exposes the difficulties associated with the task of automatic video summarization. In the context of text summarization, Mani and Maybury [219] have identified three important factors for summary creation and evaluation; Conciseness, Context and Coverage. He et al. [140] address the same issue for video summarization and identify a fourth factor; Coherence.

Video summarization started to receive interest from the research community in the mid nineties [323, 15, 306, 251, 304]. Since then, the topic has received ever increasing attention. The approaches found in the literature are extremely varied, and can be organized along a number of potential axes, such as the modalities employed to create the summary, the type of summary created (static vs dynamic), the method used for the creation (the selection process), whether the method offers generic properties or is suitable for a specific type of video. Here, we will divide the literature into 2 main categories according to the type of summary created by the method. This choice is motivated from the fact that some application may suit more closely one type of summary than the other. Having said that, it is possible to transpose a dynamic summary into a static one by performing key-frame selection. The opposite, converting a static summary into dynamic one, is also achievable by recovering shots from which key-frames were selected in order to create the video skim. We shall now report some of the approaches from the literature for dynamic and static summary creation.

2.1 Dynamic Summaries

Dynamic summaries are often referred to as video-skims. Video-skims may be seen as video previews where shots or scenes which have been classified as less important or less relevant are skipped. This type of summary has the advantage over their static counterpart in that they can combine images (video) and audio. This allows the summary to convey more information about the original content of the multimedia document.

In its most simplistic form a video-skim is created by extracting pieces of video of fixed duration at intervals uniformly distributed over the video [234]. Nam and Tawfik [81] have proposed an approach which extends the basic scheme by sub-sampling the video non-linearly. The rate depends directly on the amount of visual activity measured within the shots. Others including [243] and [248], have proposed to basically fast-forward through the video in a uniform or adaptive manner. The major drawback of such approaches is the distortion caused to the original material.

Overall, the common task of more advanced algorithms is the selection of the excerpts to retain for the summary. Obviously, this will essentially depend on the objective and the application domain of the summary.

Some view the summarization process as one where the objective is to remove redundant scenes or shots from the original document. In [71], a self similarity matrix of video features is employed to select and adjust video excerpts length. This selection process may also be addressed like a clustering problem. In [116], visual features grouped according to their similarity and excerpts which lie closest to cluster centers are used to construct the summary. Similar approaches [115, 236, 314] have extended this idea with the use of additional modalities such as audio, text, motion, etc.

In the case of domain specific methods, it might be possible to detect particular events, such as goals in a soccer game or action scenes in a movie. In [207], Lienhart et al. studied the properties generally found in a movie trailer, which resulted in a number of event detectors based on video, audio and text features. The location of events detected in the movie indicates which shots/scenes should be present in the trailer. Another event based technique was proposed by Chang et al. [49]. In this work, baseball game highlights are detected using HMM models trained on 7 different game actions.

Another class of methods achieves video summarization by looking at the evolution through time of a single or a set of features. In effect, a score (feature value) is computed and associated with each temporal video element; A shot, a frame, a caption word, a sentence, etc ... depending on the modality and the method. The selection scheme for video-skim candidates based on temporal element feature value may be threshold based, maxima based [348] or obtained in a greedy manner [319, 205]. In such approaches, the challenge is to identify the right set of features.

2.2 Static Summaries

As opposed to dynamic summaries, it is possible to present static summaries differently. The static summary may be viewed like a story-board (or a film strip), a mosaic of key-frames, a slide-show or a flowchart. Its major advantage over video-skims is the possibility to present the content with an emphasis on its importance or relevance rather then in a sequential manner.

The most basic way to create a static summary is to perform some sub-sampling on the video, at a rate based upon the number of key-frame desired [15, 306]. The major drawback of summarization through direct sub sampling is that there are little guarantees that the selected key-frames have some sort of relevance. A step toward improving the selection process is to detect content change in the video and retain key-frames from the segmented shots [323, 208, 328]. The difference between such approaches resides in the method employed to select the representative frame or frames for each shot. This may be realized in a systematic manner (i.e. by taking the first frame of each shot [323]), or competitive process over shot frames [100].

An obvious way to select candidate frames to summarize a video is to cluster features extracted from the video and identify representative key-frames from each cluster. Other approaches, view the video as a curve in a multi-dimensional space of feature where each video frame is represented as a point. In this framework proposed by DeMenthon et al. [77] the process of summarization corresponds to the selection of a set of points (frames) on the curve which retain as much as possible the general shape of the curve. In effect, polygonal approximation techniques can be employed to provide solutions. In [77], a recursive binary curve splitting algorithm is used to this end but alternative algorithms such as discrete contour evolution [43] may also be employed.

Event detection is yet another way to capture and identify important video frames. The most common attribute employed for event detection is motion. In [212], Lui et al. extract representative frames based on the analysis of the motion patterns within shots. Others [255, 44], base the selection process on characteristic motions of extracted regions from frames. Content characteristics may also be of importance in order to determine the importance of a shot. For example, knowing that a frame contains people [88] or specific objects with a given behavior [311] can take a deterministic part in the summary creation process. Approaches relying on event detection are generally too specific to deal with arbitrary videos and are therefore application domain limited.

In an attempt to obtain summaries with as much fidelity as possible to the original multimedia document, Chang et al. [48] introduced the idea of using frame with maximum frame coverage as summary representative. This idea has then been extended by Yahiaoui et al. [352] to the selection of the set of frames which are the most frequently found (or sufficiently similar to at least one frame) in excerpts of a given duration. A variant of this approach [351] has been developed for multi-episode video summarization in an attempt to exhibit the major differences between episodes of TV series. This approach ensures that the resulting summaries will have little redundancy while covering as many different aspects of the video as the number of key-frames selected.

2.3 Summary Evaluation

Evaluation of video summaries is an issue often overlooked by researchers. This is probably due to the fact that there is no standard measure to assess the quality of a summary. Moreover, the quality of a summary depends greatly on its intended purpose as well as the application domain, thus it is not possible to define a general performance measure. Furthermore, the process of summary evaluation is a highly subjective one.

The most common evaluation found in the literature [376, 362] consists in presenting results of the approach for a number of multimedia documents and providing some motivation for the selected sequences or key-frames. Some researchers go through the time-consuming process of involving users in the evaluation process. This more realistic evaluation procedure may be performed in three different manners. In the first scenario [212, 88], users or experts are asked to summarize some document in order to obtain a ground truth which can then by compared with the automatically created one. In the second scenario [83], they are asked to judge or assess of the quality of computer-generated summaries with respect to the original videos. In the last scenario [85, 350], the summaries are presented to the evaluation users along with a set of tasks or questions. The quality of the answers is then analyzed to grade the summary and therefore the underlying construction methodology.

It is nonetheless possible to define a metric in order to assess the quality of the summaries. This metric is in most cases directly derived from the fidelity factor used to perform the selection process and is therefore often biased toward the newly proposed approach [48, 376, 140]. In an effort to provide a common metric for video summarization algorithms, DeMenthon et al. [148] have proposed an automatic performance evaluation based on performance evaluation metrics used in the field database retrieval.

The review presented as introduction to this chapter on video summarization is by no mean exhaustive. For a more comprehensive review of the field we invite interested readers to have a look at the following papers [349, 359, 326].

3 A Generic Summarization Approach

In this section, we propose a new approach for the automatic summarization of audio-video sequences, based on the Maximum Recollection Principle. We show that this approach defines a sensible optimization criterion for summaries, which can be adapted to a number of situations. We first describe the approach for summaries based on video only. We then extend to the case of summaries of several videos, and for summaries using video and text jointly.

3.1 Maximum Recollection Principle

The idea for the Maximum Recollection Principle (MRP) was suggested by the situation where some people randomly zap to a TV channel, watch a few seconds and are able to recognize a movie that they have already seen. The formalization of this idea leads to the following statement:

> "The summary of a document should contain such information to maximize the probability that a user would recognize the document when exposed to an extract of the document."

This statement provides the basis for a sound framework to define optimal summaries, while leaving much flexibility in the application to various types of documents. Several arguments support the use of the MRP:

- first, it is a reasonable objective for a summary, as the "zapping example" is a very common situation,

- second, it provides a measurable criterion (probability of recognition), so that an optimal summary can be defined,
- third, it leaves open the precise definition of an extract, which information from the document is being displayed, (we have found that a random choice of the extract is a good start, but the duration of the extract is still a parameter of the summarization),
- finally, the concept of "recognition" can be implemented in a number of different ways, leading to variations which can be adapted to numerous situations (for example, different similarity measures or different document types).

The application of the Maximum Recollection Principle to video sequences gives a simple illustration of the principle. A video sequence is a sequence of images $V = I_1, I_2, \ldots I_T$. A summary is a selection of key-frames: $S = I_{s(1)}, I_{s(2)}, \ldots I_{s(k)}$ (we assume that the size k of the summary is fixed, either by the system or by the user). We suppose that a virtual user has seen the summary and is presented a random excerpt $E(r, d) = I_r, I_{r+1}, \ldots I_{r+d}$, of the video. We define by excerpt of a video a subsequence of a given duration. The user will recognize the video V if at least one of the images of the excerpt $E(r, d)$ is similar to an image in the summary. Images I and I' are similar is the value of a similarity function $sim(I, I')$ is less than a predefined threshold θ. The quality (or performance $perf(S)$) of the summary S can then be measured as the number of excerpts for which the recognition occurs. Formally,

$$perf(S) = Card\{E(r, d) : \exists i, j \ \ sim(I_{s(i)}, I_{r+j}) < \theta\}$$

(where $Card\ S$ is the number of elements of a set S).

3.2 Illustration

As an example, suppose that images have been clustered into similarity classes, so that two images are considered similar if and only if they belong to the same class. The summary is composed of a number of similarity classes (there is no need for two images of the same class in the summary). If we consider excerpts of length one, then the probability that the excerpt will be similar to an image of the summary is simply the sum of the frequencies of the class in the summary. Therefore, the optimal summary (in this simplistic case) is composed of the most frequent similarity classes. If we consider excerpts with length greater than two, then the situation is more complex, as two frequent similarity classes may often correspond to the same excerpts, and therefore be redundant in the summary. For example, assumes that the similarity classes are named A, B, C, $\ldots$, and the video is composed of images forming the sequence:

$$A\ B\ C\ .\ A\ B\ C\ .\ A\ B\ .\ A\ .$$

Table 1. Summary performance for excerpts of one frame duration (where a +
indicates a match with the excerpt starting at this position, and a - indicates no
match between summary elements and the excerpt under consideration)

Video	A	B	C	.	A	B	C	.	A	B	.	A	.	perf
$S_1 = \{A, B\}$	+	+	−	−	+	+	−	−	+	+	−	+	−	7
$S_2 = \{A, C\}$	+	−	+	−	+	−	+	−	+	−	−	+	−	6

Table 2. Summary performance for excerpts of two frames duration (where a +
indicates a match with the excerpt starting at this position, and a - indicates no
match between summary elements and the excerpt under consideration)

Video	A	B	C	.	A	B	C	.	A	B	.	A	.	perf
$S_1 = \{A, B\}$	+	+	−	+	+	+	−	+	+	+	+	+	−	10
$S_2 = \{A, C\}$	+	+	+	+	+	+	+	+	+	−	+	+	−	11

(where the period indicates another similarity class than A, B, C). With excerpts of length 1, we can draw the following performance table for summaries
$S_1 = \{A, B\}$ and $S_2 = \{A, C\}$.

The best summary is evidently the one composed with the most frequent
similarity classes, in this example, A and B, as depicted in Table 1. If excerpts
have a length of two, then excerpts overlap, so that the performance table
becomes as illustrated in Table 2.

In this case, the best summary is S_2, despite the fact that class C is less
frequent than class B. The reason is that the performance criterion will only
count one when several classes contribute to the same excerpt. Therefore, the
optimal summary will be based on a selection of classes which exhibit an
optimal mix of high frequency but also high spread throughout the video.

3.3 Maximum Recollection Principle Experiments

In order to validate our approach, we have performed a number of experiments.
Several issues are considered:

- experiment with reasonable similarity measures,
- design efficient summary construction algorithms,
- evaluate the performance of the summaries with respect to the excerpt
  length.

Visual similarity is a very difficult and complex topic. In order to define
a simple and reasonable similarity measure, we have manually labeled pairs
of images as visually similar or not, then we have compared the results of
several simple similarity measures. We have found that the measure which
provides the most coherent results with our manual labeling was based on blob
histograms. We have used this measure throughout our experiments. We have
focused our work on the efficient construction of summaries. The performance
criterion that we have defined requires a combinatorial enumeration to select

Fig. 1. An example of summary composed of the six most relevant key-frames according to our maximum recollection principle

the optimal summary. The reason is that, if we try to build the optimal summary by successively adding key-frames, the selection of a key-frame may completely rearrange the importance of the remaining key-frames. To tackle this problem, we have explored suboptimal ways of gradually constructing the summary. With our approach we build the summary in a greedy fashion by selecting the best key-frame at each step, then, when the desired size is reached, we try to replace every frame with a better frame if it exists. We found that this process was near optimal from a performance perspective, while being much faster than complete enumeration from a computational perspective. Figure 1 illustrates the results of our proposed approach on a documentary whose topic concerns principally water.

We have also experimented with the importance of the excerpt duration on the quality of summaries. In our experiments, we have used different videos: $F1$ and $F2$ are 16 minutes episodes of a TV series, H is a 50 min documentary, and C is a 45 minutes fiction. We also used the first 16 minutes $H1$ and $C1$ of H and C respectively. We have computed the performance, also referred to as coverage, of summaries composed of 6 key-frames, for excerpt durations varying from 4 to 40 seconds. The results are indicated in Fig. 2. They show that the performance increases with excerpt duration, which is obvious as there are more chances to recognize a similarity with a summary key-frame when the excerpt is longer. TV series provide a better performance, which can probably be explained by the fact that the setting of the scene is generally fixed or limited, which increases the chances of similarity. Of course, the duration of the video is also of importance, and the shorter version $H1$ and $C1$ have better performance than H and C, although still lower than the performance for TV series of the same duration.

4 Multi-Video Summarization

In this section, we describe the application of the MRP to the case of the simultaneous summarization of a set of videos. Of course, an easy solution is to construct independent summaries for each video. But in the case where

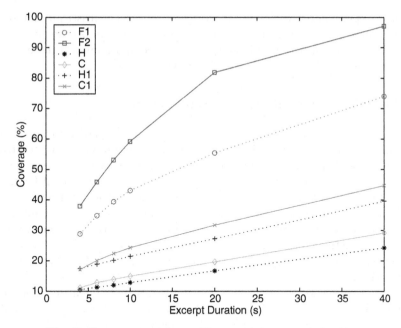

Fig. 2. Summary coverage with respect to excerpt duration

the videos are related, for example different episodes of the same TV series, this could lead to redundant information included in several summaries. In the case of related videos, we would like the summaries to contain only the information which is particular to each video, and not to contain information which is common in several or all episodes.

4.1 Definition

Assume that the user has seen the summaries of all the videos and is presented with an excerpt of an unknown video. Three cases may happen:

- if the excerpt has no similarity with any of the summaries, then the unknown video remains unknown,
- if the excerpt has some similarities with one of the summaries S, then the video is identified as the corresponding video,
- if the excerpt has some similarities with several of the summaries, then the case is ambiguous and the video cannot be uniquely identified.

According to the MRP, we should try to build a set of summaries which maximize the probability of correct answer occurring in case two, and minimize the other cases. The optimal construction of summaries becomes much more difficult in the case of multi-videos. If we try to build the summaries progressively, by adding key-frames one by one, it may happen that a key-frame which would be very good for the performance of a summary would also create many ambiguities with other previously selected key-frames.

We have experimented with several strategies to select heuristically interesting key-frames to be added, the idea being to jointly maximize the number of similarities that can be found in the current video and minimize the number of confusions that it may create in other videos.

4.2 Multi-Video Summarization Experiments

We have evaluated our algorithms for multi-video summarization on a set of 6 episodes of a TV series. The construction of the summaries is iterative, one key-frame being added to each summary in turn. The following figure (Fig. 3) shows the number of correct identifications, incorrect identifications and ambiguities in the summaries built for various excerpt durations.

We also evaluated the robustness of the summaries that are built with respect to the excerpt duration. For this, we consider the summaries built by the best method for a given excerpt duration, and we evaluate their performance for a different excerpt duration. The results are given in the figure below (Fig. 4).

These results show that, except for summaries built for a duration of 1 second, the performance remains quite stable. For example, a summary built for a duration of 4 seconds has almost the same performance on a duration of 20 seconds as the optimal summary.

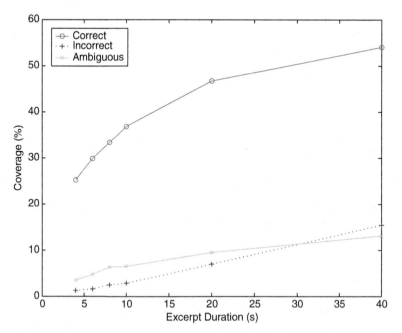

Fig. 3. Multi-Summaries evaluation with respect to excerpt duration

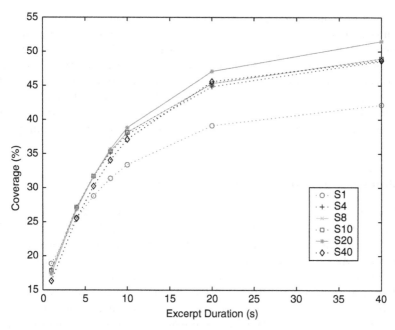

Fig. 4. Summary coverage with respect to excerpt duration

5 Joint Video and Text Summarization

In this section, we assume that the documents are composed of both video and text, the text being the synchronized transcription of the audio channel. An excerpt of the document will contain both the video and the corresponding audio. A summary will be composed of a set of key-frames and keywords. The Maximum Recollection Principle can be applied easily again in this case. An excerpt of the document will be recognized if an image of the video is similar to a key-frame of the summary, or a word in the audio is similar to a keyword of the summary. The performance of the summary is the percentage of excerpts that are recognized. One may argue about the relevance of the similarity criterion for key-words. While for images, similarity is generally a convincing argument for the similarity of the videos, this is not really true for keywords, except maybe for very rare words. In fact, keywords can be more considered as hints for similarity than complete evidence. We have explored this issue in other work (not presented here, because it deals with keywords only), for example by considering the number of documents on the Internet that are retrieved from a single keyword as an indicator of the pertinence of the keyword. Though, for joint video and text summarization, we have found that the usage of keyword similarity provides a sensible way of selecting keywords from the document. Although this issue deserves deeper investigations, we consider our current approach as a useful step in this direction. The text from the audio track is filtered (common words are removed, as in information retrieval systems, and words are stemmed). The similarity between keywords is simply the identity of the stems. The construction process of the summaries remains similar to the case of video only: the process starts with an empty summary and tries to add elements one by one. An element can be either a key-frame or a keyword. At each step, the element which best improves the performance of the summary is selected.

5.1 Joint Video and Text Summarization Experiments

We have tested our approach on several videos. We have obtained the transcription of the audio track through the caption channel. This provides us with keywords and time-code information that relates them to the corresponding key-frames. As an example, the following figure (Fig. 5) shows a summary with 10 elements for a documentary about the Amazonian forest.

Here, the construction algorithm has chosen to include one image and 9 keywords in the summary. This selection is based on the efficiency of the element in terms of similarity with the possible excerpts. We can also impose to use a predefined number of images and key-words in the summary. This may happen for example, when the summary should be displayed inside a predefined template. The following figure (Fig. 6) shows the performance of the summaries on several documents, as a function of the number of images and an excerpt duration of 20 seconds.

Fig. 5. An example of multi-modal summary

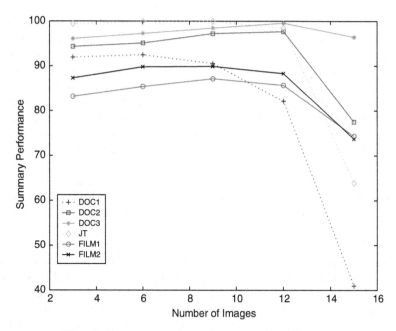

Fig. 6. Summary performance vs number of images

6 Constrained Display Summaries

During the course of a project, we got the request to produce summaries which could fit on a PDA display. Given some space limitation, we wanted to produce the best summary, in this case, this means selecting and organizing efficiently the information displayed on-screen for maximum accessibility. Therefore, we modified our construction algorithm by assigning an occupancy factor to images and key-words, for example images take five units of space, while keywords use only 1 unit. The algorithm is then able to find the best summary which respects the global occupancy constraint. At each step, the element which gets the best performance/occupancy ratio is selected. The process stops when there is no more available space to fill. In this case, the number of elements is not fixed, but chosen by the algorithm.

7 Home Video Network Interaction

In the previous sections, a generic approach to multimedia documents summarization has been proposed. We will now describe a system; developed within the European Project SPATION [225] (Services Platforms and Applications for Transparent Information management in an in-hOme Network) where video summaries are used effectively. With the ever increasing storage, networkability, and processing power of consumer electronic devices (CE-devices), developing a common platform would result in a home network with tremendous possibilities. The challenging task of storing and retrieving information in the home network was the focus of the project. Today's CE-devices are able to store hundreds of hours of video, thousands of songs, and ten thousands of photographs, etc. Solutions are required to organize and retrieve the various documents (mp3, mpeg, jpeg, etc) in a user friendly manner. To address the issue of navigation and content selection, multimedia summaries are introduced to provide an effective solution. In the SPATION project two different types of summaries are automatically computed on the devices; video trailers and visual overviews consisting of representative frames [326]. Those different types of generated summaries can be used in future CE-devices in a number of ways. The automatically generated summaries can be of great assistance to users whether they wish to locate a particular program or a scene within a program or they are trying to remember if they have already watched a program. It may also help a user to decide whether a program should be deleted, archived on DVD or simply kept on the local hard drive without having to watch it entirely. Summaries can also be extremely useful in the context of video on demand in order to decide if the entire movie should be downloaded, or simply to ease the process of choosing what to watch next out of the hundreds of hours of recorded material. In the case where summaries can be computed on the fly, when one turns a channel on and the broadcast of the program has already started if would be possible to display a preview of what has happened so far (using picture in picture view for example). Among the important functionalities related to the viewing of trailer like summaries, the ability to pause, stop, fast-forward or rewind and skip to the next one are very desirable. Indeed this would allow "zapping" through summaries as easily as zapping through broadcasts or DVD scenes. In the case of mosaic like (still picture) summaries, it is possible to refine the level of detail of summaries by increasing the number of representative images, or by selecting one of the summary thumbnails in order to access directly the corresponding scene in the original document. Another extremely useful feature relies on the fact that is it possible to download, via Bluetooth or WiFi, summaries on a mobile handheld device. The summaries may then be displayed on the PDA or *Philips' iPronto* while the user is on the move. For the SPATION demonstrator, which is depicted in Fig. 7, a *Philips iPronto* is used. This device also operates as an advanced remote control, providing access to the home CE-devices and their multimedia documents. Having the summaries available on the mobile device allows viewing shorter versions of the programs while the user is away from home. Indeed, this is also possible in the case where the program (photo, song or even video) does not fit entirely on the device permanent storage (hard drive or flash memory based storage) to replace it with a summarized version adapted to the amount of storage available.

Fig. 7. The SPATION demonstrator, presenting the user with a static summary of a broadcasted program

The SPATION demonstrator shows a realistic application for video summaries in an interactive environment. It also, proves the feasibility of such a system and clearly gives some insight into some of the functionalities of future home CE-devices.

8 Conclusion

In this chapter, we have presented our approach to video summarization, based on the Maximal Recollection Principle. Our proposal provides a criterion for the automatic evaluation of summaries, and thus allows us to define and construct optimal multimedia summaries. We have shown that it can be applied in a variety of situations, in particular it can be used to generate optimal summaries containing both video and text. Moreover, the demonstrator developed during the SPATION project exposes some of the many potential usages of multimedia summaries. Indeed, summaries will undoubtedly play an important role within future interactive multimedia devices.

Most of the existing research on video summaries has focused on criteria to detect important information in video sequences. With the current multiplicity of approaches and results, the next challenge is to establish an agreed framework for the competitive evaluation of summaries. This requires a better identification of the various usage of summaries, as well as a better understanding of the quantitative value of a summary. This is a difficult task,

Part II

Algorithms I

Building Object-based Hyperlinks in Videos: Theory and Experiments

Marc Gelgon[1] and Riad I. Hammoud[2]

[1] Laboratoire d'Informatique Nantes Atlantique (FRE CNRS 2729), INRIA Atlas project
Ecole polytechnique de l'université de Nantes, La Chantrerie, rue C.Pauc, 44306 Nantes cedex 3, France.
marc.gelgon@polytech.univ-nantes.fr
[2] Delphi Electronics and Safety, One Corporate Center, P.O. Box 9005, Kokomo, IN 46904-9005, USA.
riad.hammoud@delphi.com

1 Description of the Problem and Purpose of the Chapter

Video has a rich implicit temporal and spatial structure based on shots, events, camera and object motions, etc. To enable high level searching, browsing and navigation, this video structure needs to be made explicit [130]. In this large problem, the present chapter deals with the particular issue of object recovery, with a view to automatic creation of hyperlinks between their multiple, distant appearances in a document.

Indeed, a given object of interest almost always appear in numerous frames in a video document. Most often, these visual occurrences arise in successive frames of the video, but they may also occur in temporally disconnected frames (in the same or in different shots). Yet, by pointing at an object in an interactive video, the end-user generally wishes to interact with the abstract, general object at hand (e.g. to attach some annotation, for instance to name a person appearing in a video, or to query, in any image of a sports event, for the scoring statistics of a player) rather than its particular visual representation in the very frame he is pointing at. In other words, the user's view of the system should generally behave at a level of interpretation of the video content that considers all these visual instances as a single entity. Grouping all visual instances of a single object is the very subject of the present chapter. The paper has the following goals : (i) indicating the benefits of object tracking and matching for building rich, interactive videos, (ii) clarifying the technical

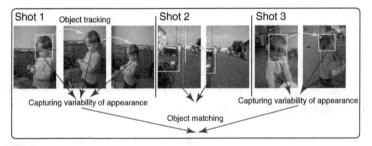

Fig. 1. Approach to structuring the video at object-level : once the video is partitioned into shots, objects of interest are detected and tracked. Then, partial groups of visual instances may be matched

problems faced by these tasks in this specific context and (iii) presenting the major frameworks that help fulfill this goal, focusing when appropriate on a particular technique.

The present paragraph exposes the general viewpoint taken in this chapter to solve the above problem. It is illustrated by Fig. 1. Gathering the various appearances of an object falls into two cases, each calling for different types of techniques. The distinction depends on whether there exists some continuity, over time, in the visual characteristics of the object (notably location, appearance etc...). Typically, this continuity will generally be present within a video shot, even if the object undergoes short occlusions. *Object tracking* is the corresponding computer vision task. Its numerous existing solutions in the literature introduce this continuity as prior knowledge. In contrast, visual instances of objects occurring in different shots will often imply a stronger variability of location and appearance between instances to be matched. We shall call *object matching* this second goal. The complete problem of grouping visual instances is more easily solved if one first determines sub-groups obtained by object tracking (intra-shot task), and matches these sub-groups in a second phase. The first phase supplies a set of collections of explicitly related visual instances. This makes the second task an original one: matching is to be carried out between collections of appearances rather than single image regions. The issues and techniques in this chapter assume that the video has already been partitioned into shots.

Tracking and matching objects during the video preparation phase grants the following advantages:

-   new means of access to the content: an object can be annotated or queried from any of its visual instances; an index of extracted objects may be presented to the user; hyperlinks can be automatically built and highlighted, that relate the temporally distant visual instances of the same physical object or person.
-   by grouping the visual instances of an object, one gathers data to better characterize it (its dynamics or visual appearance). This could be further

used to annotate the object more accurately or at a higher level of interpretation. Clearly, this characterization, on one side and object tracking & grouping, on the other side, are tightly interdependent, intricate issues.

Both object tracking and matching are major topics of interest for research contributions in video processing. While the former roots in image motion analysis and data association, the latter is tightly related to learning and recognition. In both cases, probabilistic modelling and statistical estimation found most existing solutions. The remainder of this chapter first reviews the particularities of the object detection and tracking task in the context of interactive video (Sect. 2). It then discusses object detection (Sect. 3), object tracking in Sect. 4 and object matching in Sect. 5. Section 6 finally draws conclusions.

2 Particularities of Object Detection and Tracking Applied to Building Interactive Videos

Let us first expose the main features of the general task of object detection and tracking, and examine their specificities when applied to the application at hand : what particular types of objects need to be tracked, how the video preparation and query interactions should affect the design or selection of a tracking algorithm. Depending on the degree of automation in preparing an interactive video, it may be necessary to automate the detection of objects.

The variety of objects that a tracking/matching technique encounters in building interactive videos is broad. Faces, full-size people, vehicles are popular choices, but the choice is only restricted by the user's imagination. The zones of interest may be inherently mobile or only apparently, due to camera motion. The tracking and matching tasks are clearly easier when entities of interest have low variability of appearance and exhibit smooth, rigid motion. In practice, unfortunately, objects of interest to users often undergo articulated or deformable motion, change in pose, transparency phenomena ; they may be small (e.g. sports balls). Overall, tracking and matching techniques should be ready to face challenging cases.

Though we mainly refer to interactivity in a video in the sense that the end-user benefits from enhanced browsing, interactivity generally also comes at the stage of video preparation. Indeed, video structuring cannot, in the general case, be fully automated with high reliability. An assumption here is that the automated process leaves the user with a manual correction and complementation task which is significantly less tedious that full manual preparation. Also, in complement to automatic determination of zones of interest (e.g. faces), image zones of semantic interest may have to be defined by a human hand. In principle, the system can proceed with automatic tracking and matching. In practice, depending on the trade-off between preparation reliability and time invested in this preparation, the operator may want each tracking and matching operation to be itself interactive, notably to process cases where automated algorithms failed. Further, the distinction between the video preparation phase and the interactive video usage phase may blur as technology, models and algorithms improve. As interactive video preparation become more open to the end user (e.g. not only editing text captions linked to already extracted objects, but also defining new interactive objects, which implies tracking and matching at usage time), video processing techniques should be designed to keep as wide as possible the range of visual entities we may want to make interactive. In particular, beyond classical objects that correspond to a single physical object, current work in computer vision is making possible to track and match video activities (such as a group of soccer players) [95].

Interactive video addresses the tracking task in a few works (e.g. [60, 217, 231, 322]). Most approaches quoted below were not particularly designed with interactive video in mind. Yet, they form a corpus of complementary, recognized methodologies that would be the most effective for our goal.

3 Detection of Objects

3.1 Introduction

In some contexts, for instance when browsing through surveillance videos, the sheer volume of (often streaming) data makes it desirable to automate the process of picking up objects in a first frame for further tracking. Interactivity would indeed add considerable value to contents produced by scene monitoring: most portions of videos have no interest and interesting sections can directly be focused onto. Incidentally, while object detection gains in principle to be automated for all kinds of video contents, surveillance content is generally more amenable to extracting automatically meaningful objects with reliable results than general fiction videos :

- objects that may be detected by automatic criteria (notably, motion detection) better coincide with semantic entities;
- cameras are often static, panning or zooming. In these cases, scene depth does not affect apparent motion and complex techniques to distinguish real motion from parallax are not required [179]. Robust estimation of a simple parametric dominant motion model computed over the whole image, modelling apparent motion in fact due to camera motion (see Sect. 4.3.1), followed by cancellation of this motion in the video, resets the task as a static camera one.

Object detection may broadly be categorized as either recognition from a previously learned class (e.g. faces) or, in a less supervised fashion, decision based on observations of salient features in the image, such as motion. We focus here on the latter option. While the former being covered by other chapters in this book.

3.2 Motion-Based Detection

Strictly speaking, locating mobile objects requires partitioning the scene into regions of independent motions. In practice, however, one often avoids computing the motion of these mobile objects, since this is often unreliable on a small estimation support, and one instead determines spatially disconnected regions that do not conform to the estimated background motion (if any).

At this point, differences between image intensity (or, alternatively, the normal residual flow after global motion cancellation [240]) forms the basis of observations. The subsequent decision policy (mobile or background) involves learning a statistical model of the background, for instance via a mixture of Gaussians [120], to ensure robustness to noise. It is however more tricky to model intensity variability for the alternative (motion) hypothesis - see [333] for a recent analysis of the matter.

A several difficulty in this task is that motion can only be partly observed, i.e. it is, by and large, apparent only on intensity edges that are not parallel to the motion flow. Consequently, mechanisms are commonly introduced into motion detection schemes to propagate information towards areas of lower contrast, where motion remains hidden : probabilistic modelling of the image as a Markov random field with Bayesian estimation, on one side, and variational approaches, on the other side, are two popular frameworks to recover the full extent of the object. Distinguishing meaningful displacements from noise can also rely on the temporal consistency of motion : techniques for cumulating information over time in various fashions have proved effective, especially for small or slow objects [254].

3.3 Interactive Detection

While, from the user's perspective, it is easier to define an object by its bounding box, it is generally advantageous for the subsequent tracking and matching phases to be polluted by less clutter, by having the object accurately delineated. An example recent proposal that bridges this gaps was recently disclosed in [269]. In short, the colour distributions of both the object and its background are modelled by a Gaussian mixture. They are integrated, along with contextual spatial constraints, into an energy function, which is minimized by a graph-cut technique. Interestingly, this process is interactive, i.e. the user may iteratively assist the scheme in determining the boundary, starting from sub-optimal solutions, if needed. Recent work goes further in this direction [191], by learning the shape of an object category, in order to introduce it as an elaborate prior in the energy function involved in interactive object detection.

4 Object Tracking

This section outlines possible image primitives that may represent the object to be tracked and inter-frame matching (Sect. 4.1). However, this only covers a short-term view of the problem (from one frame to the next). Section 4.2

thus sets the tracking problem into the powerful space-state framework, which incorporates sources of uncertainty and object dynamics into the solution, thereby introducing long-term considerations on tracking. Finally, we detail an object tracking technique with automatic failure detection, that was designed especially for interactive video preparation (Sect. 4.3).

4.1 Design and Matching of Object Observations

The design of the appearance model depends on the accuracy required on the boundaries of the region to be tracked. To show that the applicative nature of interactive video affects its design, let us consider two opposite cases. In hand sign gesture recognition, the background could be uniform and static (hence the tracking task simpler), but hands and fingers should be very accurately delineated in each frame. In such a situation, intensity contours are a primitive of choice for tracking. Building interactive videos with very general content (films, sports etc...) is a rather opposite challenge : determining a bounding box or ellipse on the object to be tracked is generally sufficient (at least for tracking and user interaction), but the spatio-temporal content of the scene is less predictable and subject to much clutter (notably in terms of intensity contours) and matching ambiguities.

Let us examine three major and strongly contrasted approaches for object representation and matching between two successive frames :

1. Contour-based region representations have the advantage of being light and amenable to fast tracking. These contours may be initially determined from edge maps or defined as the boundary of a region undergoing homogeneous motion [247]. Rather than tracking pixel-based contours, is it common to first map some parametric curve on the contour [180]. This constraint both regularizes the contour tracking problem and it an even lighter primitive to track. In contrast, geodesic active contours put forward in [247] are based on the geometric flow and since they model-free, they are highly effective in tracking, with accurate localization of the boundary, objects which appearance undergoes local geometric deformations (e.g. a person running). Nonetheless, such approaches come short when the background against which the object is being tracked is highly cluttered with intensity edges, which is in practice very common.

2. Color histogram representation is very classical, due to its invariance w.r.t. geometric and photometric variations (using the hue-saturation subspace) and its low computation requirements. It has regained popularity and shown very effective in state-of-the-art work ([166, 249, 378]), that makes for its drawbacks (many ambiguous matches) by applying it in conjunction with a probabilistic tracking framework (Sect. 4.2).

3. Object tracking can be formulated as the estimation of 2D inter-frame motion field over the region of interest [110]. We cover this case thoroughly in Sect. 4.3.

In the matching process, the reference representation (involved in the similarity computed in the current frame) may be extracted from the first image of the sequence, or in the most recent frame where it was found, or a more subtle combination of past representations, as proposed by the space-state framework, as the next section presents. An alternative for capturing the variety of appearances of an object can be obtain by eigen-analysis [28].

Searching for the best match of this reference representation in the image can be cast into some optimization problem of the transformation parameter space. An established approach to this is the mean-shift procedure, that searches for modes of the similarity function (e.g. Bhattacharya coefficient [67]) through an iterative procedure. The effectiveness of the optimization is based on computing the gradient of the similarity function via local kernel functions. In this matching process, peripheral pixels of the template are weighed less than central ones, to increase robustness to partial occlusions. An extension was recently presented in [135], that embeds this ideas into a Bayesian sequential estimation framework, which the next section describes.

4.2 Probabilistic Modelling of the Tracking Task

Let us consider region tracking from t to $t + 1$. With a deterministic view, once the optimal location found for the object at $t + 1$, it is considered with certainty to be the unique, optimal location. The search then proceeds in the frame at $t + 2$, in a similar fashion, and so forth. However, as already mentioned, finding correspondences in successive frames is subject to many ambiguities, especially if the object being tracked is represented by curves, interest points or the global histogram described above rather than the original bitmap pattern.

A more theoretically comprehensive and practically effective framework is Bayesian sequential estimation. A state vector contains sought information (say, location of the tracked object, also possibly its size, motion, appearance model etc...) should be designed. Tracking is then formulated as temporally recursive estimation of a time-evolving probability distribution of this state, conditional on all the observations since tracking of this object started. A full account may be found in [17]. Figure 2 provides an intuitive view of one recursion. Two models need to be defined beforehand :

- a Markovian model that describes the dynamics of this state (eqs. (1) and (2), where $x(t)$ denotes the state vector at time t, and $v(t)$ and $\mathcal{N}^v(t)$ the process noises modelling the stochastic component of state evolution). This model typically encourages smoothness of the state (for instance, of its trajectory).
- the likelihood of a hypothesized state $x(t)$ giving rise to the observed data $z(t)$ (eqs. (3) and (4), where $n(t)$ and $\mathcal{N}^n(t)$ model measurement noise). Typically, the likelihood decreases with the discrepancy between the state and the observation.

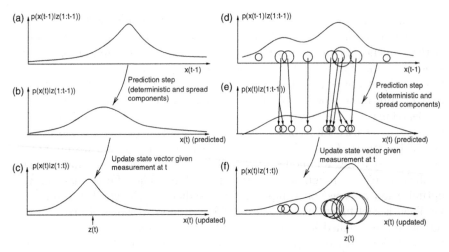

Fig. 2. Illustration of the recursive, probabilistic approach to object tracking : (a,b,c) show, in the simple case of linear evolution and observation equations and Gaussian distributions, a typical evolution for the state vector $x(t)$ during the two phases of tracking between time instants $t-1$ and t : during the prediction phase, the state density if shifted and spread, reflecting the increase in state uncertainty during this phase. Then, this prediction is updated in the light of the observation $z(t)$. Typically, the state density tightens around this observation. The same principles drive the alternative, particle-based, implementation (d,e,f) : prediction is composed of a deterministic drift of existing particles, followed by stochastic spread and hence of the corresponding density. As in (b), predicting increases uncertainty on the state. The update phase (f) is implemented by updating the particle weights according to the likelihood of new observations. Note that the state estimated at $t-1$ (d) and prediction (e) are multimodal densities, implicitly capturing two competing hypotheses, due to ambiguities in past observations. In (f), occurrence of a new observation re-inforces one of the two hypotheses, while smearing the other. This figure is after [160]

A tracking technique generally uses either eq. (1) with eq. (3), or eq. (2) with (4). The latter is in fact a simplification of the former, where temporal evolution of the state and state-to-observation relation are assumed linear, and noises are assumed Gaussian.

$$x(t) = f((x(t-1), v(t-1))) \tag{1}$$

$$x(t) = F(t) \cdot x(t-1) + \mathcal{N}^v(t-1) \tag{2}$$

$$z(t) = h((x(t), n(t))) \tag{3}$$

$$z(t) = H(t) \cdot x(t) + \mathcal{N}^n(t) \tag{4}$$

Practical computation of the recursive Bayesian estimation of $p(x(t)| z(1:t))$, i.e. the state conditional on all past information, differs in these two cases:

- in the linear/Gaussian case, a simple, closed-form solution known as Kalman filtering provides a straightforward solution. The two steps of the recursive algorithm (temporal prediction, update given a new observation) are illustrated on Fig. 2(a,b,c). However, since the posterior $p(x(t)|z(1:t))$ is Gaussian, its ability to capture uncertainty is strictly limited to some degree of spread around a single, central and necessary most probable solution.

- in contrast, the non-linear/non-Gaussian case has much richer modelling capabilities : in confusing situations (in total but temporary occlusion, or strong change in appearance), multiple hypotheses for location may be implicitly maintained via a posterior distribution with multiple modes, and propagated through time. Their respective relevances may eventually be re-considered if evidence is later supplied in favour of one of these hypotheses. However, the lack of closed-form for the posterior calls for approximation techniques. The most popular one, because of its ease of implementation, low computational cost and practical success in a wide range of applications is particle filtering [160]. With this approach, the state probability distribution is handled as a set of weighted punctual elements ('particles') drawn from the distribution. Temporal evolution of the posterior then comes down to simple computations for making these particles evolve. Figure 2(d,e,f) sketches the steps of the recursion in this case. Works in the past few years have addressed the numerous practical problems of particle filtering, such as the ability of the particle set of finite size to represent correctly the posterior when the densities are highly peaked or when the state-space is high dimensional. An alternative to particle representation was proposed in [249] in the form of a variational approximation of the posterior.

4.3 Example of a Object Tracking Technique with Failure Detection

This section focuses on a particular object tracking technique that was especially designed with the preparation of interactive videos in mind, in that the tracking is itself interactive. While a full account is provided in [110], we recall here its main features. Given a region automatically or manually defined in a first time, the scheme tracks the region until it automatically detects a tracking failure. This may happen when the object has disappeared or because of a sudden change in its appearance. When such a failure is noticed, a request is made to the human operator to redefine manually the region just before tracking failed. The remainder of this section discusses the tracking technique and the automatic failure detector.

4.3.1 Object Tracking by Robust Affine Motion Model Estimation

An essential component of the present technique is estimation of an affine motion model between successive frames, presented in [239]. With such a motion model, parameterized by $\Theta = (a_1, a_2, a_3, a_4, a_5, a_6)$, the motion vector at any pixel (x, y) within the region to be tracked is expressed as follows :

$$\omega_\Theta(x, y) = \begin{pmatrix} a_1 + a_2 \ x + a_3 \ y \\ a_4 + a_5 \ x + a_6 \ y \end{pmatrix} \tag{5}$$

Given a region determined at time t (in the form of a polygonal approximation of its boundary), the position of this region at time $t+1$ is obtain by projecting each vertex of the polygon according to the parametric motion model (5) estimated on the region between t and $t + 1$. The pixels inside the region define the support for the motion estimation. While estimating Θ is a non-linear problem, it can be effectively conducted as solving a sequence of linear problems with a least-squares procedure [239]. An important feature of the scheme is its statistical robustness, i.e. its ability to discard, in the estimation process, data that does not conform to the dominant trend in the motion model estimation. Because identification of these outliers and estimation of the dominant motion model are tightly interwoven problems, this is implemented via a Iterative Re-weighted Least Squares procedure. Robustness grants the following advantages : should the estimation support include a few pixels that do not actually belong to the region to be tracked, the reliability of the motion estimation and hence of the tracking is only very slightly affected. This situation may occur due to the approximation by a polygon, or because this polygonal model is slightly drifting away from the object, as may occur when the affine motion model is not rigorously applicable, or when there are strong, sudden changes in object appearance (pose or illumination). The technique is also robust to partial occlusions of the object being tracked.

The choice of an affine motion model is founded on an approximation of the quadratic motion model, derived from the general instantaneous motion of a planar rigid object. While this may seem a strict constraint, it can in fact capture most perspective effects that a homographic transform would fully model. As a sequence of linear problems, applied in a multi-resolution framework, its computational cost is very low, which matters in interactive tracking. One may correctly argue that its ability to handle deformable motion is limited. However, this is a price to pay for reliability of tracking in strong clutter (the technique does not assume, nor is perturbed by, intensity contours). Extensive practice has shown that the affine motion model offers a good trade-off between ability to represent geometric changes and speed and reliability in estimating the parameters of the transform, for both small and large objects. Let us point out that the reference frame number increases as pairs of successive frames are processed, hence the scheme can cope, to some extent, with strong changes of appearance that occur progressively. However, the composition of two affine transforms remains an affine transform, which limits the possible deformations in the long run.

Experimental results illustrating this tracking technique are provided in Fig. 3 for three image sequences. The first (a,b,c) and third (g,h,i,j) sequences

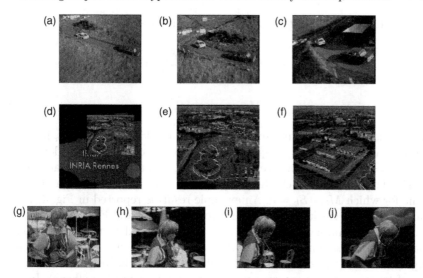

Fig. 3. Three example image sequences illlustrating object tracking based of the robust estimation of an affine motion model. Fiction excerpt sequences (a,b,c) and (g,h,i,j) also demonstrate the object matching techniques described in the next section

involve a change in pose of a non-planar object, while the region tracked in the second sequence (d,e,f) undergoes zoom and strong occlusion. In these three examples, the zones tracked were manually defined at the beginning of the sequence.

4.3.2 Automatic Failure Detection

Since the statistical robustness of the motion estimation technique determines the subset of the data that does not conform to the dominant motion model, the size of this subset, relatively to the complete estimation support, provides an indication of global model-to-data adequacy. We exploit this principle to derive a criteria for detecting tracking failure. Let $\xi_t = \frac{n_t^d}{n_t^c}$ be the ratio of the number of pixels not discarded by the robust estimator, to the size of the complete estimation support. When tracking performs correctly, ξ_t is usually close to 1. If tracking suddenly fails, the value of is variable will suddenly drop towards 0. An algorithm for detecting strong downward jumps in a signal could solve this task. However, if the polygonal model drifts away from the actual object more slowly, i.e. over several frames, this variable will take intermediate values (say, 0.8), during these frames. To detect both critical situations with a single technical solution and parameterization, we resort to a jump detection test, known as Hinkley's test [22], on the cumulative deviations from 1.

The two following quantities need to be computed:

$$S_k = \sum_{t=0}^{k} (\xi_t - 1 + \frac{\delta_{min}}{2}) \tag{6}$$

$$M_k = \max_{0 \le i \le k} S_i \tag{7}$$

The sum is computed from the frame at which tracking begins. δ_{min} indicates the minimum size of the jump one wishes to detect. A jump is detected if $M_k - S_k > \alpha$, where α is a threshold for which a value of 0.3 gave satisfactory results. The time at which the jump started may be identified as the last frame for which $M_k - S_k = 0$. An example result is reported in Fig. 4, where the object being tracked is a car (first rows in Fig. 3 and Fig. 9).

A second example (see Fig. 5) illustrates failure detection in tracking a small object (a ball in the "Mobile & Calendar" sequence ; tracking is challenging because the ball is far from planar and is undergoing a composition of translation and rotation under severe light reflection effects). Images (a,b,c,d,e,f) respectively correspond to images at time instants 1, 4, 8, 12, 14

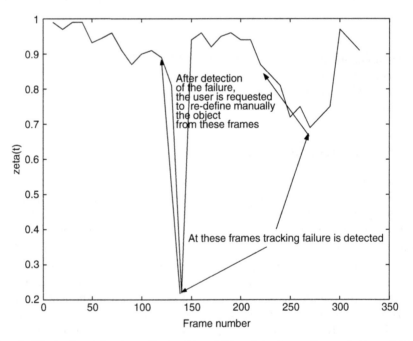

Fig. 4. Illustration of automatic tracking failure detection : the evolution of ξ_t is plotted over a 350 frame sequence. Tracking fails in two cases : in the first situation, it fails suddenly, while in the second case, the polygon defining the tracked region slowly drifts away. In both cases, the technique indicates to the interactive video preparation system the first frame it estimated ξ_t significantly departed from 1, corresponding in principle to where the user should re-define the region manually

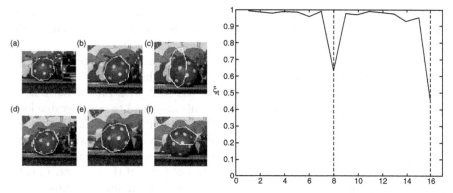

Fig. 5. "Mobile & Calendar" sequence: (left) Images (a,b,c,d,e,f) respectively correspond to images at time instants 1, 4, 8, 12, 14 and 16, and tracking failure is (correctly) detected in pictures 8 and 16. (right) The corresponding evolution of variable ξ_t shows a sudden drop when tracking fails in these two cases

and 16, and tracking failure is (correctly) detected in pictures 8 and 16. While the ball was initially automatically extracted, its boundary was manually redefined after each tracking failure.

5 Building Object-Based Hyperlinks in Video

This section describes a successful framework for building object-based hyperlinks in video sequences [124, 131, 129]. Identified objects in shots are associated to each other based on a similarity measure. We first review the issue of appearance changes of an object within the frames of a single video shot. Then we describe in details our framework to classify inter-shot objects.

Mixture of Gaussians distribution is becoming more popular in the vision community. For the problem of motion recognition, Rosales [268] evaluates the performance of different classification approaches, K-nearest neighbor, Gaussian, and Gaussian mixture, using a view-based approach for motion representation. According to the results of his experiments on eight human actions, a mixture of Gaussians could be a good model for the data distribution. McKenna [224] uses the Gaussian color mixture to track and model face classes in natural scenes (video). This work is the closest to the contribution presented in this chapter; it differs mainly by the input data which are tracked objects in our case, and in technical details like Gaussian models and the related criterion.

5.1 Appearance Changes in Video Shot

As mentioned earlier in this chapter the tracking process is performed per shot. Most objects of interest would be eventually those in motion. They are recorded in various lighting conditions, indoor and outdoor, and under different camera motion and angles. These changes in appearance make both tracking and classification of video objects very challenging.

The major questions raised here are, how to handle the large number of objects in a video sequence, and what to do with these changes in a recognition process? Perhaps a representation of the shot with one or more keyframes could be considered [132]. However, this may lead to considerable lost of relevant information, as seen in Fig. 6. This figure illustrates the changes in appearance of a tracked figure in a video shot; the changes from sunlight into shade produce a significantly bimodal distribution with two different mean colors, one for each lighting condition and tracked size of the subject. In this figure four different occurrences of a child running from sunlight into shade are showing significant changes in appearance. At the beginning, the child progressively appears and at the end of the shot he disappears. Evidently, the first occurrence will not match the middle one suggesting that an efficient recognition technique ought to consider and model somehow the temporal intra-shot variations of features.

In next section we will describe our framework for modeling such appearance variability and subsequently utilizing these models to classify occurrences of objects in different shots (inter-shots).

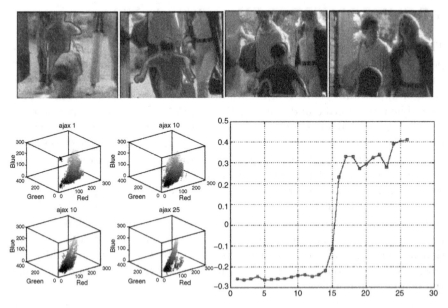

Fig. 6. Illustration of the intra-shot appearance changes in the feature space. Only frames 1, 20, 30 and 50 of the video shot are shown (left). The color histogram is computed for each appearance of the tracked subject. Then, all histograms are projected onto the first eigenvector after performing a principal components analysis

5.2 Gaussian Mixture Framework for Classifying Inter-Shots Objects

The proposed framework here encloses three major parts: (1) registration of object classes, (2) mixture modeling of registered object classes, and (3) Bayesian classification of non-registered objects. The goal here is to classify all objects in a video sequence into known object classes. The known object classes are those selected by the author of the interactive video.

5.2.1 Object Classes Registration Interface

At this step the author of the interactive video would specify the distinct objects of interest (known as "registered object classes") and the exact number of object classes. For this purpose an interface, like the one shown in Fig. 7, could be employed [130]. The author of the interactive video navigates into a list of images and selects by mouse the objects of interest. In Fig. 7 each image represent an entire video shot. It is the representative key-frame of the shot, usually the temporally median image. One click on the object would allow to register this object as a model. A registered object will be referred as "object model" in the rest of this chapter.

Fig. 7. Snapshot of the registration interface. In this example, only four object models/classes (outlined in green) were selected by the author of the interactive video. The system will then classify all remaining objects in the video sequence into these four models

The registration process is straightforward; each selected object is be assigned automatically a unique label. The author also indicates the type of the feature in the representation process. In our experiments we will validate our framework using color histograms.

5.2.2 Gaussian Mixture Modeling

Let L be the set of object models labeled by the author of the interactive video through the interface presented above. Each tracked object has many appearances in successive frames and thus all these appearances are assigned the same unique label of the object model. Let y_i be the feature vector of dimension d that characterizes the appearance object i. Let Y be the set of features vectors collected from all appearances of a single tracked object model. The distribution of Y is modeled as a joint probability density function, $f(y \mid Y, \theta)$ where θ is the set of parameters for the model f. We assume that f can be approximated as a J-component mixture of Gaussians: $f(y|\theta) = \sum_{j=1}^{J} p_j \varphi(y|\alpha)$ where the p_j's are the mixing proportions and φ is a density function parameterized by the center and the covariance matrix, $\alpha = (\mu, \Sigma)$. In the following, we denote $\theta_j = (p_j, \mu_j, \Sigma_j)$, for $j = 1, \ldots, J$ the parameter set to be estimated.

Assuming the tracked object model is in movement, it is expected to see changes in its representation, y_i, as shown in Fig. 8. These changes are more likely to be non-linear due to different types of image transformations like scale, rotation, occlusion, and global and local outdoor lighting changes. The right side of 8 shows few sample images of four selected object models. These differences between samples of the same object model translate into sparsity of the distribution (colored dots) in the first three principal components of the RGB histogram space. Each object distribution is being modeled separately by a mixture of Gaussians whose parameters are estimated as follows.

Parameters Estimation. Mixture density estimation is a missing data estimation problem to which the EM algorithm [46] can be applied. The type of Gaussian mixture model to be used (see next paragraph) has to be fixed and also the number of components in the mixture. If the number of components is one the estimation procedure is a standard computation (step M), otherwise the expectation (E) and maximization (M) steps are executed alternately until the log-likelihood of θ stabilizes or the maximum number of iterations is reached. Let $\mathbf{y} = \{y_i; \ 1 \le i \le n \ and \ y_i \in \mathbb{R}^d\}$ be the observed sample from the mixture distribution $f(y|\theta)$. We assume that the component from which each y_i arises is unknown, so that the missing data are the labels c_i $(i = 1, \ldots, n)$. We have $c_i = j$ if and only if j is the mixture component from which y_i arises. Let $\mathbf{c} = (c_1, \ldots, c_n)$ denote the missing data, $\mathbf{c} \in B^n$, where $B = \{1, \ldots, J\}$. The complete sample is $\mathbf{x} = (x_1, \ldots, x_n)$ with $x_i = (y_i, c_i)$. The complete log-likelihood is $L(\theta, \mathbf{x}) = \sum_{i=1}^{n} \log \left\{ \sum_{j=1}^{J} p_j \varphi(x_i|\mu_j, \Sigma_j) \right\}$.

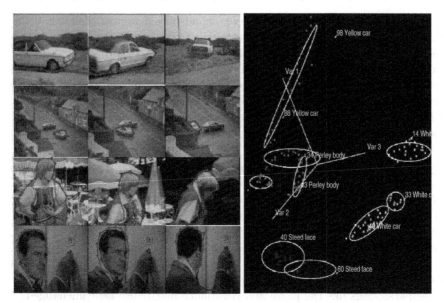

Fig. 8. Capturing the intra-shot variability of the four tracked objects (left side) by Gaussian Mixture Models in the color histogram space (right side). Each appearance of a tracked object is being represented by a color histogram which is then projected in the first three eigen vectors. The distribution of each object is modeled by a mixture of Gaussian with dynamic number of components.

More details on the EM algorithm could be found in [78, 46]. Initialization of the clusters is done randomly. In order to limit dependence on the initial position, the algorithm is run several times (10 times in our experiments) and the best solution is kept.

Gaussian models. Gaussian mixtures are sufficiently general to model arbitrarily complex, non-linear distribution accurately given enough data [224]. When the data is limited, the method should be constrained to provide better conditioning for the estimation. The various possible constraints on the covariance parameters of a Gaussian mixture (e.g. all classes have the same covariance matrix, an identity covariance matrix, ..), defines 14 models. In our experiments we have utilized the following seven models derived from the three general families of covariance forms: $M_1 = \sigma_j^2 I$ and $M_7 = \sigma^2 I$ the simplest model from the spherical family (I is the identity matrix); $M_2 = \sigma_j^2 \text{Diag}(a_1, \ldots, a_d)$ and $M_3 = \sigma_j^2 \text{Diag}(a_1^j, \ldots, a_d^j)$ from the diagonal family where $|\text{Diag}(a_1^j, \ldots, a_d^j)| = 1$ with unknown $a_1^j, \ldots, a_d^j$; M_4 from the general family which assumes that all components have the same orientation and identical ellipsoidal shapes; M_6 from the general family which assumes that all covariance matrices have the same volume; Finally, M_5 is the most complex model with no restrictions. More details on the estimation process could be found in [46].

Model choice criterion. To avoid a hand-picked number of modes of the Gaussian mixture, the Bayesian Information Criterion (BIC) [288] is used to determine the best probability density representation (appropriate Gaussian model and number of components). It is an approach based on a measure that determines the best balance between the number of parameters used and the performance achieved in classification. It minimizes the following criterion: $BIC(M) = -2L_M + Q_M ln(n)$ where L_M is the maximized log-likelihood of the model M and Q_M is its number of free parameters.

5.2.3 Classification of Non-Registered Object Occurrences

After object registration and parameter estimation, the task at hand now is to automatically classify any remaining object occurrence in the large video sequence into these object models. First, the Gaussian components of all object models are brought together into one global Gaussian mixture density of **K** components, where $\mathbf{K} = \sum_{\ell=1}^{\mathbf{L}} J_\ell$ and J_ℓ is the number of components of the ℓth object model Ω_ℓ. Only the proportion parameter is re-estimated while other parameters like the mean and covariance matrix are kept unchanged.

The posteriori probability $t_{i\ell}(\theta)$ that object occurrence y_i belongs to class Ω_ℓ is given by:

$$Prob(\Omega_\ell) = t_{i\ell}(\theta) = \sum_{j=1}^{J_\ell} Prob(Z = \ell_j \mid y_i, \theta). \tag{8}$$

Note that the object model ℓ is a mixture of J_ℓ Gaussian where J is estimated during modeling phase using BIC criteria. The a posterior probability that y_i belongs to a Gaussian component of the global Gaussian mixture of class Ω_ℓ is given by:

$$Prob(Z = k \mid y_i, \theta) = \frac{p_k \varphi(y_i \mid \mu_k, \Sigma_k)}{\sum\limits_{j=1}^{K} p_j \varphi(y_i \mid \mu_j, \Sigma_j)}. \tag{9}$$

The object occurrence y_i is then classified into the object model with the highest a posteriori probability $t_{i\ell}(\theta)$. This rule is well known by Maximum A Posteriori (MAP).

Finally, the MAP rule is applied on each object occurrence in the video. Given that a video sequence is segmented into multiple shots, one could utilize the tracking information in order to make the classification process more robust to some outliers. The final step of our classification strategy consists of assigning all object occurrences of the same tracked object into the object model (class) with the highest number of votes from y_i with $1 \leq i \leq n_r$, and n_r is the total number of occurrences. Despite this effort one would expect that such classification process would remain challenging especially when the same object appears substantially different in distinct shots of the movie film.

Fig. 9. Subset of tracked object tests

5.3 Framework for Experimental Validation

Using a challenging pre-segmented video sequence of 51 tracked objects and over 1000 frames, we obtained a correct classification rate of 86% (see Table 1). Figures 8 and 9 shows few image samples of both registered object models and non-registered (test) objects. In this experiment, 15 object models were selected and registered using our interface presented above. This experiment has been repeated with different 15 object models without any remarkable changes in the performance [131]. The maximum number of permitted Gaussian components, $MaxNbC$, was ranged from 1 to 4. The BIC criteria is being used to determine the appropriate number of Gaussian components as well the best fitting Gaussian model (among seven competitive models, see Sect. 5.2.2). We have employed a simple feature vector, the color histogram, to represent objects in feature space. The histogram approach is well known as an attractive method for object recognition [112, 102] because of its simplicity, speed and robustness. The RGB space is quantized into 64 colors [355]. Then, the Principal Component Analysis (PCA) was applied on the entire set of initial data in order to reduce their dimensionality (d_E in Table 1). This step is quit important toward overcoming the curse of dimensionality, especially when the number of samples of an object model is low and insufficient to estimate efficiently an optimal Gaussian mixture model, and also toward speeding up the estimation of Gaussian parameters.

We noticed that classification rate has increased by at least 5% when the tracking data ($track.\%$) is utilized compared to independent classification ($indi.\%$) of individual occurrences (see Sect. 5.2.3). This increase is not

Table 1. Test results with Gaussian mixture (mix.), Key-frame and mean-histogram methods.

Meth.	feat. hist.	d_E	$MaxNbC$ or dist.	Total indi. %	Total trac. %
mix.	RGB	10	1	73.65	82.99
mix.	RGB	10	2	79.10	86.10
mix.	**RGB**	**10**	**3**	**81.50**	**86.30**
mix.	RGB	10	4	71.09	73.87
key	RGB	64	χ^2	45.16	45.91
key	**RGB**	**10**	d_e	**42.90**	**44.50**
mean	**RGB**	**64**	χ^2	**43.09**	**45.00**
mean	RGB	10	d_e	47.90	45.50

significant. It is mainly due to the dramatic changes in appearance of the same object in two shots and that, in some cases, it is likely to have the highest number of occurrences miss-classified. In our test video, it seems that the cut between shots is very fast leaving for each shot an average of 50 frames only. Also, most objects are recorded outside, from different angles including airborne, and their appearance is quit variable. Evidently, the employed color histogram is not robust to all of these changes, and therefore it plays a role in miss-classifying objects.

The benefits of the classification framework could be seen when comparing the obtained results to other methods such as matching based on representative frames [238] and average-feature. A key-frame is selected randomly to represent the entire shot, and thus each object model has only one representative appearance. The average-feature method [369] consists of computing the mean of all feature vectors of all object model's occurrences. Two metrics are used in the matching procedure, namely, χ^2-test and Euclidean distance d_e. Table 1 summarizes the obtained results by the three different methods. An increase by 40% of the correct classification rate is noticed when our proposed classification framework is employed.

6 Conclusion and Perspectives

Hyperlinks between objects of interest are a main appealing feature of interactive video. Automating their generation is based on a challenging combination of computer vision tasks : automatic detection of objects (optional), intra-shot tracking and matching across shots. This chapter has reviewed the specificities of their application to interactive video, the main technical issues encountered in this task and solid, recent approaches from the literature. Techniques for tracking and matching, specifically designed for interactive video, are detailed and illustrated on real data.

Clearly, such a system is only as strong as its components, and its reliability will improve according to advances in elementary technologies. Besides this, nonetheless, there are perspectives to make the tracking and matching steps better benefit from one another. From tracking in a space-state framework to matching, the full state posterior could be exploited by the matching phase to propagate uncertainty and multiple hypotheses. Conversely, the variability of appearance learned in the matching phase could provide valuable information to the tracking scheme. Finally, since detection, tracking and matching typically follow an inference-decision scheme, one could introduce asymmetric costs, in the decision phase, to the various errors, reflecting the costs of manual correction for the various mistakes that an automatic scheme makes (e.g. cancelling erroneous objects may be easier than manually defining new ones), or what errors may or may not be tolerated by an end user.

Acknowledgments

I would like to thank my former sponsor and institution, Alcatel Alshthom Research and INRIA Rhone-Alpes for their support. Also, we are so grateful to the INA (french national institute of audio-visual) for providing us the video material used in our benchmarking.

Real Time Object Tracking in Video Sequences

Yunqiang Chen[1] and Yong Rui[2]

[1] Siemens Corporate Research, 755 College Rd Ease, Princeton, NJ 08540, USA.
Yunqiang.Chen@siemens.com
[2] Microsoft Research, One Microsoft Way, Redmond, WA 98052, USA.
yongrui@microsoft.com

1 Introduction

Reliable object tracking in complex visual environment is of great importance. In addition to its applications in human computer interaction [159, 162], tele-conferencing [334, 338], and visual surveillance [187], it can also play an important role in the interactive multimedia/video framework. As pointed in [276], an unstructured video clip can be organized into key frames, shots and scenes. If we can reliably track an object, it can help detect when the object is of interest (e.g., bigger size), and can therefore extract more meaningful key frames. Also, object tracking can provide useful information for shot boundary detection. For example, a sudden loss of tracked objects may indicate a new shot. Furthermore, object segmentation and tracking can provide high-level semantic information about the video (e.g. the objects in the video and their motion patterns or interactions), which can be important for classifying similar shots to form a scene or for content-based query in video databases [51].

To start object tracking, usually the trackers need to be initialized by an external module. For example, a human operator can select an object of interest and let the tracking begin. For a more intelligent tracking system, an automatic object detection module can be used to initialize the tracking. Automatic object detection algorithms are usually trained based on a set of images of typical object appearances at different states and viewed from different angles, e.g. the multi-view face detector [202] or hand posture recognition [344].

Once initiated, the tracking algorithms will conduct tracking based on the high correlations of the object motion, shape or appearance between consecutive video frames. Unfortunately, robust and efficient object tracking is still an open research problem. Two of the major challenges are:

1. Visual measurements for tracking objects are not always reliable. To discriminate objects from background clutter, various image cues have been

proposed. Object contour, face template or color distributions are used in [161, 65, 66, 343] respectively. In complex environments, none of the above features are robust enough individually. More and more researchers are therefore resorting to multiple visual cues [27, 260, 162, 335]. The major difficulty for multicue object tracking is, however, how to effectively integrate the cues in a principled way.

2. Tracking objects in nonlinear dynamic systems is not easy. While we know that Kalman filter provides an elegant solution for linear systems, in the real world, the states of the objects are usually nonlinearly related to the measurements through observation models. For example, the contour points of an ellipsoid object are nonlinearly related to the object's position and orientation.

The Hidden Markov Model (HMM) [258] provides a potential tool to solve the first difficulty. It can integrate multiple visual cues by expanding the observation vectors and encode the spatial constraints in the state transition probabilities. Optimal contour can be obtained by the efficient Viterbi algorithm. However, extending the HMM structure from 1D time series to 2D imagery data is challenging. A pseudo-2D-HMM (embedded HMM) has been proposed for character recognition [192], face recognition [235], and template matching [265]. A two-level HMM is defined, where super states are used to model the horizontal dimension and embedded states are used to model the vertical dimension. However, this approach requires a large number of parameters to be trained. Instead, we propose a new type of HMM that can probabilistically integrate multiple cues under various spatial constraints, including global shape prior, contour smoothness constraint, and region smoothness constraint. Parametric shape is used to model object contour. Multiple visual cues (e.g., edge and color) are collected along the normal lines of the predicted contour (see Fig. 1(a)). We define the HMM states to be the contour point location on each line. This representation allows us to formulate the 2D (in image plane) contour tracking into an easier-to-solve 1D problem.

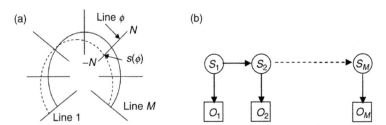

Fig. 1. The new contour model: *(a) The contour in 2D image space: The solid curve is the predicted contour. The dashed curve is the true contour. We want to find the $s(\phi)$ which is the index of the true contour point on the ϕth normal line $\phi \in [1, M]$; (b) the Markovian assumption in our contour model. Note the HMM states are in the spatial domain*

The transition probabilities represent the prior constraint on the contour points, such as the contour smoothness. For better transition probabilities in addition to the traditional contour smoothness [8], we further develop an efficient joint probability matching term to encode the contexture information in the neighborhood around the contour. The new transition probabilities are estimated based on a well-defined energy minimization procedure and enforce both the contour smoothness and region smoothness constraints. This formulation enables us to achieve better contour detection in cluttered environments [55].

To address the second difficulty, sequential Monte Carlo methods [159] and the extended Kalman filter (EKF) [174] have been proposed in the past. However, the computational complexity of sequential Monte Carlo methods grows exponentially with the state space dimension. The EKF can be more efficient but it only approximates the Taylor expansion to the first order. It therefore does not work well in the systems where nonlinearity is severe. Furthermore, it can be difficult to compute the Jacobians and Hessians of a nonlinear system as required by the EKF. We utilize a new type of nonlinear filter, the Unscented Kalman filter (UKF). Unlike the widely used EKF, the UKF approximates the Taylor expansion up to the second order (third for Gaussian priors). Furthermore, the UKF does not need to explicitly calculate the Jacobians or Hessians. Therefore, it not only outperforms the EKF in accuracy (second order vs. first order approximation), but also is computationally efficient. Its performance has been demonstrated in many applications [176, 330, 57].

By letting the HMM integrate multiple cues under spatial constraints and the UKF handle nonlinear system dynamics in the temporal domain, the combined HMM-UKF provides a powerful parametric contour tracking framework. Centered at this new framework are the energy-minimization processes used in the HMM's transition probability estimation and the UKF's object state estimation. At each time frame t, we use the HMM to obtain the best object contour. The results are then passed to the UKF as the measurements to estimate the object parameters based on the object dynamics. The UKF provides an MMSE estimation of the object states. Furthermore, a robust online training process is proposed to adapt the object properties over time and make it possible to track objects in non-stationary environments.

The rest of this chapter is organized as follows. In Sect. 2, we present a new HMM framework that achieves more robust contour detection by integrating multiple visual cues and spatial constraints. In Sect. 3, we define a more comprehensive energy term that incorporates not only the contour smoothness constraint but also the region smoothness constraint based on an efficient joint probabilistic matching. In Sect. 4, we extend our tracking into the temporal domain and use the UKF to estimate object states in the nonlinear system. An adaptive learning process is further developed to update the object properties through time to handle object appearance changes.

Experiments and comparisons in Sect. 5 show that the proposed HMM-UKF achieves robust tracking results. Concluding remarks are given in Sect. 6.

2 Contour Tracking Using HMM

For tracking nonrigid objects, active contour models have been proved to be powerful tools [181, 320, 250]. It uses an energy minimization procedure to obtain the best contour, where the total energy consists of an internal energy term for contour smoothness and an external energy term for edge likelihood. For real-time tracking, it is imperative to have an efficient energy minimization method to find the optimal contour. In traditional active contour methods, the optimization procedure is not very efficient due to the recursive contour refinement procedure [8, 80]. Considering the aperture effect, where only the deformations along the normal lines of the contour can be detected, we can restrict the contour searching to a set of normal lines only (see Fig. 1(a)). In this way, we convert the 2D searching problem into a simpler 1D problem. To define the 1D contour model, let $\phi = 1, ..., M$, be the index of the normal lines and $\lambda = -N, ..., N$, be the index of pixels along a normal line and $\rho_\phi(\lambda)$ denote the image intensity or color at pixel λ on line ϕ:

$$\rho_\phi(\lambda) = I(x_{\lambda\phi}, y_{\lambda\phi}) \tag{1}$$

where $(x_{\lambda\phi}, y_{\lambda\phi})$ is the corresponding image coordinate of the pixel λ on the ϕth normal line. $I(x_{\lambda\phi}, y_{\lambda\phi})$ is the image intensity or color at $(x_{\lambda\phi}, y_{\lambda\phi})$.

Each normal line has $2N + 1$ pixels, which are indexed from $-N$ to N. The center of each normal line is placed on the predicted contour position and indexed as 0. If the object had moved exactly as predicted, the detected contour points on all normal lines would have been at the center, i.e., $s(\phi) = 0, \forall \phi \in [1, M]$. In reality, however, the object can change its motion and we need to find the true contour point $s(\phi)$ based on the pixel intensities and various spatial constraints. Note that instead of representing the contour by a 2D image coordinate, we can now represent the contour by a 1D function $s(\phi), \phi = 1, ..., M$.

To detect the contour points accurately, different cues (e.g., edge and color) and prior constraints (e.g., contour smoothness constraint) can be integrated by an HMM. The hidden states of the HMM are the true contour points on all the normal lines, denoted as $\mathbf{s} = \{s_1, ..., s_\phi, ..., s_M\}$. The observations of the HMM, $\mathbf{O} = \{O_1, ..., O_\phi, ..., O_M\}$, are collected along all the normal lines. An HMM is specified by the observation model $P(O_\phi|s_\phi)$ and the transition probability $p(s_\phi|s_{\phi-1})$. Given current state s_ϕ, the current observation O_ϕ is independent of the previous state $s_{\phi-1}$ and the previous observation $O_{\phi-1}$. Because of the Markovian property, we have $p(s_\phi|s_1, s_2, ..., s_{\phi-1}) = p(s_\phi|s_{\phi-1})$, which is illustrated in Fig. 1(b).

In Sect. 2.1, we give detailed descriptions on how to incorporate multiple cues in the observation model. A simple state transition is then discussed in

Sect. 2.2 based on the contour smoothness constraint. In Sect. 2.3, the Viterbi algorithm [258] for finding the optimal contour is described. Unlike other contour models [109] that also resort to the dynamic programming to obtain the global optimal contour, the HMM offers an elegant way to integrate multiple visual cues and a probabilistic training formula (shown in Sect. 4.2) to adapt itself in dynamic environments.

2.1 Observation Likelihood of Multiple Cues

In the HMM, the observation on line ϕ (represented as O_ϕ) can include multiple cues. We describe the observation model based on color (i.e., $\rho_\phi(\lambda), \lambda \in [-N, N]$) and edge detection (i.e., $\mathbf{z}_\phi$) along the line in this section.

First, the observation likelihood based on the edge detection ($\mathbf{z}_\phi$) can be derived similar to [161]. Because of noise and image clutter, there can be multiple edges along each normal line. Let J be the number of detected edges, we have $\mathbf{z}_\phi = (z_1, z_2, ..., z_J)$. Of the J edges, at most one is the true contour. We can therefore define $J + 1$ hypotheses:

$$
\begin{aligned}
H_0 &= \{e_j = F : j = 1, ..., J\} \\
H_j &= \{e_j = T, e_k = F : k = 1, ..., J, k \neq j\}
\end{aligned}
\tag{2}
$$

where $e_j = T$ means that the jth edge is the true contour, and $e_j = F$ otherwise. Hypothesis H_0 therefore means the true contour is not detected by the edge detection. With the assumption that the clutter is a Poisson process along the line with spatial density γ and the true target measurement is normally distributed with standard deviation σ_z, we can obtain the edge likelihood model as follows:

$$
p(\mathbf{z}_\phi | s_\phi = \lambda_\phi) \propto 1 + \frac{1}{\sqrt{2\pi}\sigma_z q \gamma} \sum_{m=1}^{J} \exp(-\frac{(z_m - \lambda_\phi)^2}{2\sigma_z^2})
\tag{3}
$$

where q is the prior probability of hypothesis H_0. A typical edge-based observation along one normal line is shown in Fig. 2. Multiple peaks appear due to clutter.

To reduce the clutter, HMM can easily integrate other cues, such as color histogram of the foreground (FG) and background (BG). Let v be the color, $p(v|FG)$ and $p(v|BG)$ represent the color distribution for the FG and BG respectively. If $s_\phi = \lambda_\phi$ is the contour point on line ϕ, we know that the segment $[-N, s_\phi]$ of line ϕ is on the FG and the segment $[s_\phi + 1, N]$ is on the BG. Combining the edge likelihood model and the color histogram of the FG/BG, we have the following multicue observation likelihood model:

$$
P(O_\phi | s_\phi) = p(\mathbf{z}_\phi | s_\phi) \cdot \prod_{i=-N}^{s_\phi} P(v = \rho_\phi(i) | FG) \cdot \prod_{i=s_\phi+1}^{N} P(v = \rho_\phi(i) | BG)
\tag{4}
$$

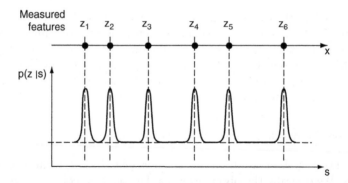

Fig. 2. One-dimensional edge-based observation $p(\mathbf{z}_\phi|s_\phi)$: $z_1, z_2, ..., z_J$ are the detected edges along one normal line and cause a multiple peak edge-based likelihood model

Other cues can also be integrated in a similar way. As we will show in Sect. 4.2, our proposed HMM framework also allows us to update the foreground/background color model in a probabilistic way over time.

2.2 Computing Transition Probabilities

In addition to the observation model discussed in the previous subsection, another important component in the HMM is the transition probability. It determines how one state transits to another. In this subsection, we use the standard contour smoothness constraint to derive the transition probability. We defer a more complete smoothness constraint based on region properties to Sect. 3.

The contour smoothness constraint can be encoded in transition probability. To enforce the contour smoothness constraint in the HMM, the constraint needs to be represented in a causal form. In Fig. 1(a), we can see that when the normal lines are dense (30 lines in our experiments), the true contour points on adjacent normal lines tend to have similar amounts of displacement from the predicted contour position (indexed as 0 on each normal line). This constraint is causal and can be captured by transition probabilities $p(s_\phi|s_{\phi-1})$ defined as follows:

$$p(s_\phi|s_{\phi-1}) = c \cdot e^{-(s_\phi-s_{\phi-1})^2/\sigma_s^2} \tag{5}$$

where c is a normalization constant and σ_s regulates the smoothness of the contour. This transition probability penalizes sudden changes of the adjacent contour points, resulting in a smoother contour.

2.3 Best Contour Searching by Viterbi Algorithm

Given the observation sequence $\mathbf{O} = \{O_\phi, \phi \in [1, M]\}$ and the transition probabilities $a_{i,j} = p(s_{\phi+1} = j|s_\phi = i)$, the best contour can be found by finding the most likely state sequence $\mathbf{s}^*$. This can be efficiently accomplished by the Viterbi algorithm [258]:

$$\mathbf{s}^* = \arg\max_{\mathbf{s}} P(\mathbf{s}|\mathbf{O}) = \arg\max_{\mathbf{s}} P(\mathbf{s}, \mathbf{O}) \tag{6}$$

Let us define

$$V(\phi, \lambda) = max_{s_{\phi-1}} P(O_\phi, s_{\phi-1}, s_\phi = \lambda) \tag{7}$$

Based on the Markovian assumption, it can be recursively computed as follows:

$$V(\phi, \lambda) = P(O_\phi|s_\phi = \lambda) \cdot \max_j P(s_\phi = \lambda|s_{\phi-1} = j)V(j, \phi - 1) \tag{8}$$

$$j^*(\phi, \lambda) = P(O_\phi|s_\phi = \lambda) \cdot \arg\max_j P(s_\phi = \lambda|s_{\phi-1} = j)V(j, \phi - 1) \tag{9}$$

with the initialization $V(1, \lambda) = \max_{s_1} P(O_1|s_1)P(s_1)$, where the initial state probabilities $P(s_1) = \frac{1}{2N+1}, s_1 \in [-N, N]$. The term $j^*(\phi, \lambda)$ records the "best previous state" from state λ at line ϕ. We therefore obtain at the end of the sequence $\max_{\mathbf{s}} P(\mathbf{O}, \mathbf{s}) = \max_\lambda V(M, \lambda)$. The optimal state sequence $\mathbf{s}^*$ can be obtained by back tracking j^*, starting from $s_M^* = \arg\max_\lambda V(M, \lambda)$, with $s_{\phi-1}^* = j^*(s_\phi^*, \phi)$. The computational cost of the Viterbi algorithm is $O(M \cdot (2N + 1))$. Unlike traditional active contour model [8, 80], this method can give us the optimal contour without recursively searching the 2D image plane. The best state sequence $\mathbf{s}^* = \{s_1^*, ..., s_M^*\}$ will be used in the Unscented Kalman Filter (UKF) to calculate the innovations and estimate the contour parameter in Sect. 4.1.

3 Improving Transition Probabilities

The transition probability is one of the most important components in an HMM. It encodes the spatial constraints between the neighboring contour points. In Sect. 2.2, we derive a simplified way of computing transition probabilities based on the contour smoothness constraint. Even though simple, it only considers the contour points themselves and ignores all the other pixels on the normal lines, which can be dangerous especially when the clutter also has smooth contour and is close to the tracked objects (e.g., the dark rectangle in Fig. 3(a)). To estimate the contour transition robustly, we should consider all detected edges jointly similar to the Joint probability data association (JPDAF) in [20]. We introduce joint probabilistic matching (JPM) into the HMM for calculating more accurate transition probabilities. With contexture information, the new transition probabilities are more robust. An efficient optimization algorithm based on dynamic programming is developed to calculate the JPM term in real time.

3.1 Encoding Region Smoothness Constraint Using JPM

Since the true contour can be any pixel on the normal line, we have to estimate the transition between the pixels on the neighboring normal lines. Let s_ϕ and $s_{\phi+1}$ be the contour points on line ϕ and line $\phi+1$, respectively. These two contour points segment the two lines into foreground (FG) segments and background (BG) segments. If the object is opaque, the edges on FG cannot be matched to BG. To exploit this constraint, we need to track the transitions of all the pixels on FG/BG together. That is, it is not a matching of contour points only, but rather a matching of the whole neighboring normal lines. The transition probabilities based on this new matching paradigm enforce not only the contour smoothness but also region smoothness constraint and are therefore more accurate and robust to clutter.

Let $E^F(i,j)$ and $E^B(i,j)$ be the matching errors of the neighboring foreground segments (i.e., segment $[-N, i]$ on line ϕ and $[-N, j]$ on line $\phi+1$) and background segments (i.e., segment $[i+1, N]$ on line ϕ and $[j+1, N]$ on line $\phi+1$), respectively. Let $\delta(i) = j$ specify that pixel i on line ϕ should be matched to pixel j on line $\phi+1$ and $\delta()$ is monotonic (i.e., $\delta(i-1) <= \delta(i)$). Then, the matching cost is defined as:

$$E^F(i,j) = min_\delta \sum_{k=-N}^{i} ||\rho_\phi(k) - \rho_{\phi+1}(\delta(k))||_2, \quad \delta(k) \in [-N, j] \qquad (10)$$

$$E^B(i,j) = min_\delta \sum_{k=i+1}^{N} ||\rho_\phi(k) - \rho_{\phi+1}(\delta(k))||_2, \quad \delta(k) \in [j+1, N] \qquad (11)$$

A more accurate transition probability can then be estimated based on the matching cost (compare with Eq. (5)):

$$log(p(s_2|s_1)) = E^F(s_1, s_2) + E^B(s_1, s_2) + (s_2 - s_1)^2/\sigma_s^2 \qquad (12)$$

The importance of the new matching cost can be illustrated by a synthesized image in Fig. 3. There are two regions where the grey region is the object to track and the darker rectangle is a background object. There are two adjacent normal lines shown in the figure, i.e., line 1 and line 2. Points 'a' and 'b' are detected edge points on line 1. Similarly, points 'c' and 'd' are detected edge points on line 2. Our goal is to find the true contour points on these two normal lines. The pixel intensities along these two lines are shown in Fig. 3(b). They are similar to each other except for some distortions. Based on the contour smoothness constraint only, the contour from 'a' to 'c' and the contour from 'b' to 'c' have almost the same transition probabilities because $|a - c| \approx |b - c|$. However, if we consider all the pixels on the normal lines together, we can see that 'ac' is not a good contour candidate because 'b' and 'd' are now on foreground and background respectively and they have no matching on the neighboring lines. The contour candidates 'ad' and 'bc' are better because they segment the two normal lines into matching FG/BG

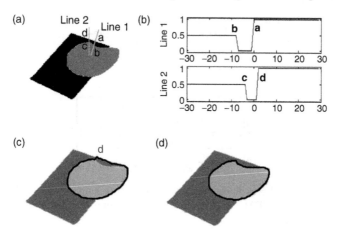

Fig. 3. Illustration of the JPM: *(a) Synthesized image with the grey object to track. (b) The observation on normal line 1 and 2. (c) Based on traditional contour smoothness only, the detection is distracted by strong continuous edges on the background. (d) With the JPM, the contour is correctly detected*

(The best choice between these two should be further decided based on Viterbi algorithm in 2.3.)

The comparison between traditional smoothness constraint and JPM based smoothness constraint is shown in Fig. 3(c) and (d). Without joint matching terms, the contour is distracted by the strong edge of the background clutter in Fig. 3(c). In Fig. 3(d), the matching cost has large penalty for the contour to jump to background clutter and then jump back. Hence we obtain the correct contour.

Unlike the uniform statistic region model in [58], our matching term is more relaxed. The object can have multiple regions (e.g., the side view of human head with face and hair as in the experiments), each of which has continuous boundaries. To illustrate this, another test is shown in Fig. 4 which has different intensity regions in the foreground. We can see the difference between Figs. 3 and 4: the observations on line 2 are not the same. There is no segment 'cd'. No matter we match 'c' to 'a' or 'b', the other edge will have no matching part. Therefore, the matching cost is the same for matching 'a' to 'c' or 'b' to 'c'. The algorithm favors the result in Fig. 4(b) because it is smoother.

3.2 Efficient Matching by Dynamic Programming

To ensure real-time performance, we propose an efficient algorithm to calculate the JPM. There are $(2N + 1)^2$ possible state transitions between the neighboring normal lines. We propose an efficient dynamic programming algorithm to calculate all $(2N + 1)^2$ matching probabilities with $2 \cdot (2N + 1)^2$

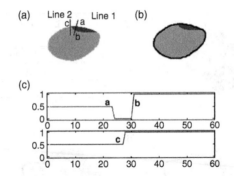

Fig. 4. Foreground object with multiple regions: (a) The synthesized image. (b) Contour tracking with JPM. (c) The observation on line 1 and line 2

computational cost. Given observation on lines 1 and 2, the calculation of the matching probabilities can be explained in the following recursive equation:

$$E^F(i,j) = \min(E^F(i-1,j)+d, E^F(i,j-1)+d, E^F(i-1,j-1))+e(i,j) \quad (13)$$

where $e(.,.)$ is the cost of matching two pixels. $E^F(i,j)$ is the minimal matching cost between segment $[-N,i]$ on line 1 and segment $[-N,j]$ on line 2. We start from $E^F(-N,j) = E^F(i,-N) = 0$, where $i,j \in [-N,N]$ and use the above recursion to obtain the matching cost $E^F(i,j)$ from $i = -N$ to N and $j = -N$ to N. A similar process is used to calculate $E^B(i,j)$, but starting from $E^B(i,N) = E^B(N,j) = 0$ and propagate to $E^B(-N,-N)$. With all the matching cost, the state transition probabilities can be computed as in Eq.(12) and contour detection can be accomplished by the Viterbi algorithm described in Sect. 2.3.

4 Combining HMM with Unscented Kalman Filter

The previous two sections have described the contour detection for each individual frame. In this section, we will extend it to tracking in video sequences. In the temporal domain, there are also object dynamics constraints, based on which we can combine the contour detection results from successive frames to achieve more robust tracking. Kalman filter is a typical solution for linear systems [9]. But in visual tracking systems, the object parameters (e.g., head position and orientation) usually cannot be measured directly but are only nonlinearly related to the observations (e.g., contour points of the object). In Sect. 4.1, we utilize the Unscented Kalman filter (UKF) to handle the non-linearity in observation model and object dynamics. Then, in Sect. 4.2, we explain how to probabilistically update the object properties (e.g., color histogram) using Balm-Welch algorithm in dynamic scenes. A complete diagram for the HMM-UKF tracking framework is given in Sect. 4.3.

4.1 UKF for Nonlinear Dynamic Systems

Let $X_{0:t}$ and $Y_{0:t}$ represent the state trajectory and observation history of a system from time 0 to time t. Tracking is the process of estimating system's current states, based on the past and current observations, i.e., $p(X_t|X_{t-1}, Y_{0:t})$. For different applications, state X_t and observation Y_t can be different entities. For example, X_t can be the position and orientation of a head, and Y_t can be the contour of the head. Based on the Markovian assumption, it can be modeled by the following equations:

$$p(X_t|X_{t-1}) : X_t = f(X_{t-1}, m_t) \tag{14}$$

$$p(Y_t|X_t) : Y_t = h(X_t, n_t) \tag{15}$$

where Eq. (14) is the object dynamics, which specifies object motion. Equation (15) is the object observation model which relates the object states to the observations. The terms m_t and n_t are process noise and observation noise, respectively. The goal of object tracking can be treated so as to minimize the mean square error:

$$\hat{X}_{t|t} = \arg\min_{\hat{X}} ||\hat{X}_{t|t} - \hat{E}(X_t|Y_{0:t})||^2 \tag{16}$$

If $f()$ and $h()$ are linear functions and if Gaussian distribution is assumed for X_t, m_t, and n_t , it has an analytical solution which is the well-known Kalman filter [9]:

$$\hat{X}_{t|t} = \bar{X}_{t|t-1} + K_t \cdot (Y_t - \bar{Y}_{t|t-1}) \tag{17}$$

where K_t is the Kalman gain, $\bar{X}_{t|t-1}$ is the predicted states, and $\bar{Y}_{t|t-1}$ is the predicted measurement based on object dynamics. Unfortunately, real-world applications seldom satisfy the Kalman filter's requirements.

As a concrete example, we use head tracking (see Sect. 5) to illustrate the object state, system dynamics model, and measurements. Specifically, a human's head is modeled by a five-element ellipse, $\Theta = [c_x, c_y, \alpha, \beta, \Phi]^T$, where (c_x, c_y) is the center of the ellipse, α and β are the lengths of the major and minor axes of the ellipse, and Φ is the orientation of the ellipse. The object states at frame t include the shape parameters and the velocity:

$$X_t = \begin{bmatrix} \Theta_t \\ \dot{\Theta}_t \end{bmatrix} \tag{18}$$

The object states are changing over time according to the object dynamics. We adopt the Langevin process [334] to model the head dynamics:

$$\bar{X}_{t|t-1} = f(X_{t-1}, m_t) = \begin{bmatrix} 1 & \tau \\ 0 & a \end{bmatrix} \begin{bmatrix} \Theta_{t-1} \\ \dot{\Theta}_{t-1} \end{bmatrix} + \begin{bmatrix} 0 \\ b \end{bmatrix} m_t \tag{19}$$

where $a = \exp(-\beta_\theta \tau)$, $b = \bar{v}\sqrt{1 - a^2}$. β_θ is the rate constant, m is a thermal excitation process drawn from Gaussian distribution $N(0, Q)$, τ is the discretization time step, and $\bar{v}$ is the steady-state root-mean-square velocity.

The object states cannot be measured directly but are only nonlinearly related to the observed object contour. As shown in Fig. 1(a), we restrict the contour detection along the normal lines of the predicted object position. It can be detected efficiently using the JPM-HMM described in Sects. 2 and 3. The detected best contour point on each normal line in Eq. (6) is used to form the 1D measurement vector:

$$Y_t = \mathbf{s}^* = [s_1^*, s_2^*, ..., s_M^*]^T \tag{20}$$

The physical meaning of $s_\phi(\phi \in [1, M])$ is the distance from the true contour point to the center (predicted contour point) of the normal line ϕ. It is nonlinearly related to the object state X_t by the observation model $h()$ in Eq. (15). For a given normal line ϕ, represented by its center point $[x_\phi, y_\phi]$ and its angle θ_ϕ, the s_ϕ can be calculated as the intersection of the ellipse $\Theta_t = [c_x, c_y, \alpha, \beta, \Phi]$ and the normal line ϕ. We first perform a coordinate translation and rotation to make the ellipse upright, centered at the origin. In the new coordinate, the center of the normal line $([x_\phi', y_\phi']^T)$ and the angle θ_ϕ' is therefore:

$$\begin{bmatrix} x_\phi' \\ y_\phi' \end{bmatrix} = \begin{bmatrix} \cos \Phi & \sin \Phi \\ -\sin \Phi & \cos \Phi \end{bmatrix} \begin{bmatrix} x_\phi - c_x \\ y_\phi - c_y \end{bmatrix} \tag{21}$$

$$\theta_\phi' = \theta_\phi - \Phi \tag{22}$$

The distance between the intersection and the center of the normal line (i.e. s_ϕ) does not change in the new coordinate and can be calculated by solving the following equation:

$$\frac{(x_\phi' + s_\phi \cos \theta_\phi')^2}{\alpha^2} + \frac{(y_\phi' + s_\phi \sin \theta_\phi')^2}{\beta^2} = 1 \tag{23}$$

We then have

$$\hat{s}_\phi = h(X_t, n_t) = n_t + \frac{-\left[\frac{x_\phi' \cos \theta_\phi'}{\alpha^2} + \frac{y_\phi' \sin \theta_\phi'}{\beta^2}\right]}{\left[\frac{\cos^2 \theta_\phi'}{\alpha^2} + \frac{\sin^2 \theta_\phi'}{\beta^2}\right]} + \frac{\sqrt{\left[\frac{x_\phi' \cos \theta_\phi'}{\alpha^2} + \frac{y_\phi' \sin \theta_\phi'}{\beta^2}\right]^2 - \left[\frac{\cos^2 \theta_\phi'}{\alpha^2} + \frac{\sin^2 \theta_\phi'}{\beta^2}\right]\left[\frac{x_\phi'^2}{\alpha^2} + \frac{y_\phi'^2}{\beta^2} - 1\right]}}{\left[\frac{\cos^2 \theta_\phi'}{\alpha^2} + \frac{\sin^2 \theta_\phi'}{\beta^2}\right]} \tag{24}$$

From the above equation, we can see that the relationship between the measurements and the object states (i.e., $h(.)$) is highly nonlinear. This high nonlinearity prevents us from effectively using the EKF [9], as it requires explicit calculation of the complicated Jacobians. The UKF provides a better alternative [176, 330]. Rather than linearizing the system, the UKF generates sigma points (i.e. hypotheses in the object state space, see Eq. (26)) and applies the true system models to these sigma points, and then uses the results to estimate the posterior mean and covariance. Compared with the EKF, the UKF does not need to explicitly calculate the Jacobians or Hessians. It therefore not only outperforms the EKF in accuracy (second-order approximation vs. first-order approximation), but also is computationally efficient. Its performance has been demonstrated in many applications [176, 330].

The UKF is implemented using the Unscented Transformation [176] by expanding the state space to include the noise component: $x_t^a = [x_t^T m_t^T n_t^T]^T$. Let $N_a = N_x + N_m + N_n$ be the dimension of the expanded state space, where N_m and N_n are the dimensions of noise m_t and n_t, and let Q and R be the covariance for noise m_t and n_t; the UKF can be summarized as follows [176, 330]:

1. Initialization:

$$\bar{x}_0^a = [\bar{X}_0^T\ 0\ 0]^T, \quad P_0^a = \begin{bmatrix} P_0 & 0 & 0 \\ 0 & Q & 0 \\ 0 & 0 & R \end{bmatrix} \tag{25}$$

2. Iterate for each time instance t:
   a) Generate $2N_a + 1$ scaled symmetric sigma points based on the state mean and variance:

   $$\mathcal{X}_{t-1}^a = [\bar{x}_{t-1}^a\ \bar{x}_{t-1}^a + \sqrt{(N_a + \lambda)P_{t-1}^a}\ \bar{x}_{t-1}^a - \sqrt{(N_a + \lambda)P_{t-1}^a}] \tag{26}$$

   b) Pass all the generated sigma points through the system dynamics in Eq. (19) and observation model in Eqs. (21 - 24). Compute the mean and covariance of the predicted object states $\bar{X}_{t|t-1}$ and predicted contour position $\bar{Y}_{t|t-1}$ using the sigma points (with corresponding weights W_i):

   $$W_0^{(m)} = \lambda/(L + \lambda) \tag{27}$$

   $$W_0^{(c)} = \lambda/(L + \lambda) + (1 - \alpha^2 + \beta) \tag{28}$$

   $$W_i^{(m)} = W_i^{(c)} = 1/\{2(L + \lambda)\} \qquad i = 1, ..., 2L \tag{29}$$

   $$\mathcal{X}_{t|t-1}^x = f(\mathcal{X}_{t-1}^x, \mathcal{X}_{t-1}^v), \quad \bar{X}_{t|t-1} = \sum_{i=0}^{2N_a} W_i^{(m)} \mathcal{X}_{i,t|t-1}^x \tag{30}$$

   $$\mathcal{Y}_{t|t-1} = h(\mathcal{X}_{t|t-1}^x, \mathcal{X}_{t-1}^n), \quad \bar{Y}_{t|t-1} = \sum_{i=0}^{2N_a} W_i^{(m)} \mathcal{Y}_{i,t|t-1}^x \tag{31}$$

   $$P_{t|t-1} = \sum_{i=0}^{2N_a} W_i^{(c)} [\mathcal{X}_{i,t|t-1}^x - \bar{X}_{t|t-1}][\mathcal{X}_{i,t|t-1}^x - \bar{X}_{t|t-1}]^T \tag{32}$$

where $\lambda = \alpha^2(L + \kappa) - L$ is a scaling parameter. α determines the spread of the sigma points around the mean. κ is a secondary scaling parameter which we set to 0, and β is used to incorporate prior knowledge of the distribution of x (we set $\beta = 2$, optimal for Gaussian distributions).

c) Calculate the Kalman gain K_t and compute the innovation $Y_t - \bar{Y}_{t|t-1}$ based on the measurement $Y_t = [s_1^*, s_2^*, ..., s_M^*]^T$ on the current frame (obtained using the JPM-HMM described in Sects. 2 and 3). Update the mean and covariance of the states:

$$P_{Y_t Y_t} = \sum_{i=0}^{2N_a} W_i^{(c)} [\mathcal{Y}_{i,t|t-1} - \bar{Y}_{t|t-1}][\mathcal{Y}_{i,t|t-1} - \bar{Y}_{t|t-1}]^T \tag{33}$$

$$P_{X_t Y_t} = \sum_{i=0}^{2N_a} W_i^{(c)} [\mathcal{X}_{i,t|t-1}^x - \bar{X}_{t|t-1}][\mathcal{Y}_{i,t|t-1}^x - \bar{Y}_{t|t-1}]^T \tag{34}$$

$$K_t = P_{X_t Y_t} P_{Y_t Y_t}^{-1} \tag{35}$$

$$\bar{X}_t = \bar{X}_{t|t-1} + K_t(Y_t - \bar{Y}_{t|t-1}), \quad P_t = P_{t|t-1} - K_t P_{Y_t Y_t} K_t^T \tag{36}$$

UKF has been proved to be more accurate and efficient than EKF in many applications [177, 332, 331, 189]. In a recent paper [193], EKF and UKF are reported to have similar accuracy while EKF is more efficient for quaternion tracking. This result seems to be contradicting other researcher's results. However, if we take a closer look, it is clear that [193] addresses a very special case. First, its quaternion dynamics is "quasi-linear" which gives EKF biased benefits. Second, their system dynamics is an integral function and demands more computation for evaluation. Third, they speed up the EKF by making approximations in calculating the Jacobian matrix, which can be dangerous when not used appropriately.

In our system, we have highly non-linear system (see Eq. (24)) and do not have the unusual integral function. Our observation is consistent with other researchers [177, 332, 331, 189], i.e., UKF is more accurate and efficient. Let's next briefly exam the computation counts for both UKF and EKF in our case. Remember that we use ellipse to model the object and Langevin process to model the dynamics and use 30 normal lines to detect the contour (i.e. $N_x = 2 \times 5 = 10$, $N_m = 5$ and $N_n = M = 30$). To calculate Jacobian matrix for EKF, we need to compute $\partial \hat{s}_\phi / \partial x_i$ for $\phi \in [1, M]$ and $i = 1, ..., N_x$ (i.e. $N_x \times M = 10 * 30$ terms). For UKF, it is very straight forward and only need to propagate $2 \times (10 + 5 + 30) + 1$ sigma points through the Eq (19) and Eqs (21 - 24), which is much easier than calculating $\partial \hat{s}_\phi / \partial x_i$.

4.2 Online Learning of Observation Models

In a dynamic environment, both the objects and background may gradually change their appearances. An online training is therefore necessary to adapt the observation likelihood models dynamically. A naive way is to completely trust the tracking results from the UKF at the current frame, and average all the pixels inside and outside the contour to obtain the new foreground/background color model for tracking on next frame. However, if error occurs on the current frame, this procedure may adapt the model in the wrong way. Fortunately, like the traditional HMM model, the JPM-HMM also allows us to adapt the observation models in a probabilistic way.

Instead of completely trusting the contour obtained at frame $t - 1$, we can make a soft decision of how to update the observation models by using the forward-backward algorithm. The "forward probability distribution" is defined as follows:

$$\alpha_\phi(s) = p(O_1, O_2, ..., O_\phi, s_\phi = s) \tag{37}$$

which can be computed efficiently using the following recursion:

$$\alpha_1(s) = p(s_1 = s)p(O_1|s_1 = s) \tag{38}$$

$$\alpha_{\phi+1}(s) = \left[\sum_u \alpha_\phi(u)a_{u,s}\right] p(O_{\phi+1}|s_{\phi+1} = s) \tag{39}$$

Similarly, the "backward probability distribution" is:

$$\beta_\phi(s) = p(O_{\phi+1}, O_{\phi+2}, ..., O_M, s_\phi = s) \tag{40}$$

which can be computed efficiently using the following recursion:

$$\beta_M(s) = 1 \tag{41}$$

$$\beta_\phi(s) = \sum_u a_{s,u}p(O_{\phi+1}|s_{\phi+1} = u)\beta_{\phi+1}(u) \tag{42}$$

After computing the forward and backward probabilities, we can compute the probability of having the true contour point at s on line ϕ:

$$P(s_\phi = s|\mathbf{O}) = \frac{\alpha_\phi(s)\beta_\phi(s)}{\sum_u \alpha_\phi(u)\beta_\phi(u)}, \quad s \in [-N, N] \tag{43}$$

Based on these probabilities, the probability of pixel λ_ϕ being on the foreground can be computed by integrating $P(s_\phi = s|\mathbf{O})$ where $s \in [\lambda_\phi, N]$ (i.e. the true contour point on line ϕ is to the outside of pixel λ_ϕ).

$$P(\lambda_\phi \in FG) = \sum_{s=\lambda_\phi}^{s=N} p(s_\phi = s|\mathbf{O}) \tag{44}$$

$$P(\lambda_\phi \in BG) = \sum_{s=-N}^{s=\lambda_\phi-1} p(s_\phi = s|\mathbf{O}) = 1 - P(\lambda_\phi \in FG) \tag{45}$$

This probability gives us a robust way to weigh different pixels during the observation models adaptation. The color distribution model (i.e. the color histogram) of foreground and background can be estimated based on the probability of each pixel belonging to foreground or background. The update equations are as follows:

$$p(v|BG) = \frac{\sum_{s=-N}^{N} P(s \in BG) \cdot (O_\phi(s) == v)}{\sum_{s=-N}^{N} P(s \in BG)}$$

$$p(v|FG) = \frac{\sum_{s=-N}^{N} P(s \in FG) \cdot (O_\phi(s) == v)}{\sum_{s=-N}^{N} P(s \in FG)} \tag{46}$$

where the more confidently classified pixels will contribute more to the color model and the less confidently classified pixels will contribute less.

The adapted color models follow the changing color distributions during the tracking. It is worth noting that the color close to the contour is more important than that of the middle in helping us locate the contour. So we only use the pixels on the normal lines to update the color models. The updated color models are then plugged back into Eq. (4) during the contour searching on next frame.

4.3 Complete Tracking Algorithm

The complete HMM-UKF tracking procedure is summarized as follows (also see Fig. 5):

1. **Prediction:** Predict where the object will be in the current frame t based on the tracking results in the previous frame $t - 1$ and the object's dynamics in Eq. (19). Observations are collected along a set of normal lines of the predicted contour.
2. **Contour tracking:**
   a) **Observation likelihood:** Evaluate the likelihood of being the true contour for every pixel on the normal line ϕ $p(O_\phi|s_\phi = \lambda_\phi)$ based on edge detection and the FG/BG color models using Eq. (4).
   b) **Transition probabilities:** Evaluate the state transition probabilities based on JPM as shown in Eq. (12). Efficient dynamic programming solution is explained in Sect. 3.2.

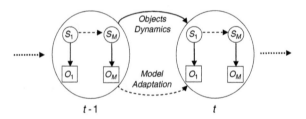

Fig. 5. The tracking diagram

c) **Contour points detection**: With previously computed observation likelihood and the transition probability matrix, the best contour can be found efficiently using the Viterbi algorithm described in Sect. 2.3.

d) **Object states estimation**: Use the detected contour points as measurements for the UKF and update the object states in the current frame as shown in Sect. 4.1.

3. **Model adaptation**: Use the forward-backward algorithm to estimate a soft classification of each pixel (to foreground and background) on the normal lines and update the color model of foreground and background based on Eq. (46). Update the velocity of the objects (e.g., translation, rotation and scaling). Go to step 1 when new frame arrives.

To begin this tracking procedure, a separate initialization module is needed. This can be done either manually or by change detection [318]. We manually initialize on the first frame of each sequence in the experiments.

5 Experiments

To validate the efficiency and robustness of the proposed framework, we apply it in various real-world video sequences captured by a pan/tilt/zoom video camera in a typical office environment. The sequences simulate various tracking conditions, including appearance changes, quick movement, out-of-plane rotation, shape deformation, background clutter, camera pan/tilt/zoom, and partial occlusion. In the experiments, we use the parametric ellipse $\Theta_t = [c_x, c_y, \alpha, \beta, \Phi]$ to model human heads and we adopt the Langevin process as the object dynamics, as discussed in Sect. 4. For all the testing sequences, we use the same algorithm configuration, e.g., object dynamics, state transition probability calculation. For each ellipse, 30 normal lines are used, i.e., $M = 30$. Each line has 21 observation locations, i.e., $N = 10$. The tracking algorithm is implemented in C++ on Windows XP platform. No attempt is made on code optimization and the current system can run at 60 frames/sec comfortably on a standard Dell PIII 1G machine. It is worth noting that the proposed HMM-UKF tracking framework is a general framework for parametric contour tracking. While we apply it in head tracking in this section, it is applicable to many other domains.

To demonstrate the strength of the proposed multicue energy-driven tracking paradigm, we conduct three sets of comparisons in this section. First, we compare our multicue tracker with the CamShift tracker in OpenCV package [156], which tracks objects using color histogram only. While CamShift represents one of the best single cue trackers available, it is still easily distracted when similar color is presented in the background. On the other hand, by using multiple cues, our tracker achieves much more robust tracking result. In the second comparison, we show the superiority of the energy-driven

JPM-HMM over the plain HMM model. In the third experiment, we compare the performance of the UKF against a least-mean-square (LMS) ellipse fitting plus Kalman filtering. By exploiting object dynamics in the nonlinear system, the UKF outperforms the Kalman filter and survives the partial occlusions. In all the experiments, the tracker is hand initialized at the first frame of a sequence and begin tracking through the sequence.

5.1 Multiple Cues vs. Single Cue

Sequence A, shown in Fig. 6, is in a cluttered office environment with 400 frames captured at 30 frames/second. Note that the sharp edges of the door and the blinds impose great challenges to contour-only tracking algorithms. The color of the door is very similar to that of a human face and causes great difficulties to color-only tracking algorithms. The comparison between CamShift tracker and the proposed HMM-UKF is shown in Fig. 6. The tracking results of the CamShift tracker are shown in the top row and the results of our approach are shown in the bottom row.

The CamShift tracker relies on an object color histogram and runs in real time. It is quite robust when the object color remains the same and there is no similar color presented in the background. Because it only models the foreground color and no information about background color is used, it is easily distracted when patches of similar color appear in the background (see frames in the top row in Fig. 6). Also, there is no principled way to update the color histogram of the object. When the person turns her head and the face is not visible (Frame 187), CamShift tracker loses track.

On the other hand, the proposed HMM-UKF probabilistically integrates multiple cues into the HMM model. Furthermore, both foreground and background color are modeled and probabilistically updated during tracking. It effectively handles object appearance change when the person turns her head

Frame 15   Frame 25   Frame 87   Frame 187

Frame 15   Frame 25   Frame 87   Frame 187

Fig. 6. Comparison between CamShift tracker and proposed approach on Sequence A: The top row is the result from CamShift and the bottom row is the result from the proposed approach

(Frame 187) and is robust to similar color in the background (e.g., Frames 15 and 25). When the color-based likelihood cannot reliably discriminate the object from background, edge-based likelihood will be weighted more during the contour detection (as shown in Eq. (4)). The HMM-UKF therefore provides much more robust tracking performance.

Sequence B, shown in Fig. 7, is a 200 frame sequence captured at 30 frames/second. It is a cluttered environment with multiple people's presence. There are both appearance and lighting changes of the person's head. Similar to the results in sequence A, the CamShift tracker is distracted by other people, while the HMM-UKF is quite robust (see Fig. 7).

5.2 JPM-HMM vs. Plain HMM

As we have described in Sect. 3, the traditional contour smoothness constraint only takes into account the contour points but ignores the other pixels around the boundaries. This can be dangerous, especially in a cluttered environment. With the new comprehensive JPM based transition probabilities defined in Eq. (12), the new HMM model is more robust to background clutter. We compare the JPM-HMM against the plain HMM model (using the contour smoothness constraint in Eq. (5)).

The comparison for sequence A is shown in Fig. 8, where the top row is the result from the plain HMM and the bottom row is the result from the JPM-HMM. In the plain HMM, tracking is distracted by the strong edges on the background when the foreground boundary does not have high contrast. The error gradually increases from frame 39 to frame 45. Because JPM models a more comprehensive spatial constraint (i.e., both the contour smoothness and region smoothness), the JPM-HMM is much more robust against sharp edges on the background.

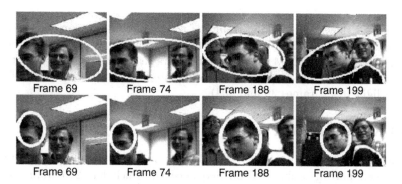

Fig. 7. Comparison between CamShift tracker and proposed approach on Sequence B: The top row is the result from CamShift and the bottom row is the result from the proposed approach

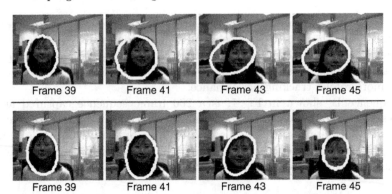

Fig. 8. Comparison between the plain HMM and JPM-HMM on Sequence A: The top row is the result of the plain HMM and the bottom row is the result of the JPM-HMM

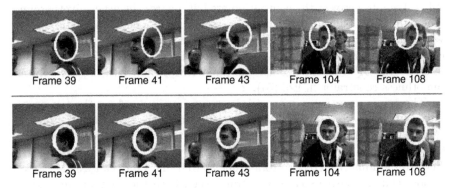

Fig. 9. Comparison between the plain HMM and JPM-HMM on Sequence B: The top row is the result of the plain HMM and the bottom row is the result of the JPM-HMM

Figure 9 shows the tracking results of the two approaches on Sequence B, where there are both appearance and lighting changes in the person's head. The proposed JPM-HMM still successfully tracks the target throughout the sequence, while the plain HMM fails, distracted by sharp boundaries of the ceiling light (e.g., frames 39 and 104).

5.3 UKF vs. Kalman Filter

In this subsection, we evaluate HMM-UKF's ability to handle large distractions, e.g., partial occlusions, on Sequence C, which has 225 frames captured at 30 frames/second (see Fig. 10). Note that the observation model in Eqs. (21 - 24) of the tracking system is highly nonlinear. It is therefore quite difficult to calculate the Jacobians as required by the EKF. An ad hoc method to get around this situation is not to model the nonlinearity explicitly.

Fig. 10. Comparison between Kalman filter and the UKF on Sequence C: The top row is the result from ellipse fitting plus Kalman filter and the bottom row is the result from the UKF

For example, one can first obtain the best contour positions and then use least-mean-square (LMS) to fit a parametric ellipse, followed by a Kalman filtering. Although simple, it does not take advantage of all the information available. The UKF, on the other hand, not only exploits system dynamics in Eq. (19) and observation model in Eq. (24), but also avoids computing the complicated Jacobians. The comparison between the UKF and Kalman filter is shown in Fig. 10.

As we can see in Fig. 10, the person's head is occluded by his hand. Neither the color nor the edge detection can reliably obtain the true contour points. The Kalman filter fails at frame 150 when large distraction occurs, because the fitting and Kalman filter are separate and could not fully utilize the dynamics constraint. By explicitly modeling the nonlinearity and the object's dynamics, the UKF, on the other hand, successfully handles the partial occlusion.

6 Conclusion

Object tracking algorithms provide high-level semantic information about the objects in the videos and their motion trajectories and interactions, which can be very helpful in understanding the videos or classifying/querying the videos in the multimedia database. Initialization is necessary to start the tracking process. It can be done either manually or by an automatic object detection module (e.g. face detection [202]).

In the interactive video/multimedia framework, it is important that the tracking modules can be initialized easily. Some tracking methods require strict and precise initialization. For example, many color based tracking methods (e.g., [66, 27]) require a typical object color histogram. In [27], a side view of the human head is used to train a typical color model of both skin color and hair color to track the human head with out-of-plane rotation.

Our proposed HMM-UKF framework can be initialized by a rough bounding box indicating the object position and then adapt itself to the changing appearance or environments, which allows it to be easily integrated with external face detector or manual initialization. In terms of the algorithm itself, several important features distinguish the proposed HMM-UKF from other approaches:

- *Multiple cues:* By expanding HMM's observation vectors, the proposed HMM-UKF framework easily incorporates multiple cues collected along object contour (e.g., edge likelihood, and foreground/background color) in a principled way.
- *Comprehensive JPM energy term:* Instead of using just the contour smoothness constraint, our JPM contour energy term incorporates the more sophisticated contexture information. This results in increased robustness in detecting object contours.
- *Nonlinearity:* The UKF's ability to approximate nonlinear systems up to the second order (third in Gaussian priors) allows us to deal with real-world complications such as large distractions (Fig. 10).
- *Online adaptation:* The HMM-UKF framework also offers an online learning process to adapt to changing environments (see Sect. 4.2).
- *Real-time performance:* The algorithm runs very efficiently. With no specific optimization of the C++ code, the algorithm can achieve real-time performance when tracking two or three objects simultaneously on a Intel P3 1GHz CPU.

To future improve the tracking results, it is possible to combined the HMM modeling with particle filters [56] to handle non-Gaussian systems and maintain multiple hypotheses during object tracking.

On Film Character Retrieval in Feature-Length Films

Ognjen Arandjelović[1] and Andrew Zisserman[2]

[1] Department of Engineering, University of Cambridge, UK.
`oa214@cam.ac.uk`
[2] Department of Engineering, University of Oxford, UK.
`az@robots.ox.ac.uk`

1 Introduction

The problem of automatic face recognition (AFR) concerns matching a detected (roughly localized) face against a database of known faces with associated identities. This task, although very intuitive to humans and despite the vast amounts of research behind it, still poses a significant challenge to computer-based methods. For reviews of the literature and commercial state-of-the-art see [21, 372] and [252, 253]. Much AFR research has concentrated on the user authentication paradigm (e.g. [10, 30, 183]). In contrast, we consider the content-based multimedia retrieval setup: our aim is to retrieve, and rank by confidence, film shots based on the presence of specific actors. A query to the system consists of the user choosing the person of interest in one or more keyframes. Possible applications include:

1. **DVD browsing:** Current DVD technology allows users to quickly jump to the chosen part of a film using an on-screen index. However, the available locations are predefined. AFR technology could allow the user to rapidly browse scenes by formulating queries based on the presence of specific actors.
2. **Content-based web search:** Many web search engines have very popular image search features (e.g. `http://www.google.co.uk/imghp`). Currently, the search is performed based on the keywords that appear in picture filenames or in the surrounding web page content. Face recognition can make the retrieval much more accurate by focusing on the content of *images*.

We proceed from the *face detection* stage, assuming localized faces. Face detection technology is fairly mature and a number of reliable face detectors have been built, see [144, 226, 285, 336]. We use a local implementation of the method of Schneiderman and Kanade [285] and consider a face to be correctly

Fig. 1. Face detection: Automatically detected faces in a typical frame from the feature-length film "Groundhog day". The background is cluttered, pose, expression and illumination very variable

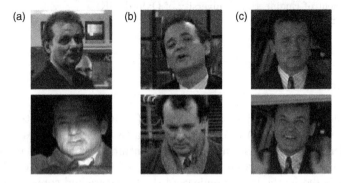

Fig. 2. Face recognition difficulties: The effects of imaging conditions – illumination (a), pose (b) and expression (c) – on the appearance of a face are dramatic and present the main difficulty to AFR

detected if both eyes and the mouth are visible, see Fig. 1. In a typical feature-length film, using every 10th frame, we obtain 2000-5000 face detections which result from a cast of 10-20 primary and secondary characters (see Sect. 3).

Problem Challenges

A number of factors other than identity influence the way a face appears in an image. Lighting conditions, and especially light angle, drastically change the appearance of a face [5]. Facial expressions, including closed or partially closed eyes, also complicate the problem, just as head pose does. Partial occlusions, be they artefacts in front of a face or resulting from hair style change, or growing a beard or moustache also cause problems. Figure 2 depicts the appearance of a face under various illumination conditions, head poses and facial expressions. Films therefore provide an especially challenging, realistic working environment for face recognition algorithms.

Method Overview

Our approach consists of computing a numerical value, a *distance*, expressing the degree of belief that two face images belong to the same person. Low distance, ideally zero, signifies that images are of the same person, whilst a large one signifies that they are of different people.

The method involves computing a series of transformations of the original image, each aimed at removing the effects of a particular extrinsic imaging factor. The end result is a *signature image* of a person, which depends mainly on the person's identity (and expression, see Sect. 2.5.2) and can be readily classified. The preprocessing stages of our algorithm are summarized in Fig. 3 and Table 1.

1.1 Previous Work

Most previous work on AFR focuses on user authentication applications, few authors addressing it in a setup similar to ours. Fitzgibbon and Zisserman [103] investigated face clustering in feature films, though without explicitly using facial features for registration. Berg et al. [25] consider the problem of clustering detected frontal faces extracted from web news pages. In a similar manner to us, affine registration with an underlying SVM-based facial feature detector is used for face rectification. The classification is then performed in a Kernel PCA space using combined image and contextual text-based features. The problem we consider is more difficult in two respects: (i) the variation in imaging conditions in films is typically greater than in newspaper photographs, and (ii) we do not use any type of information other than visual cues (i.e. no text). The difference in the difficulty is apparent by comparing the examples in [25] with those used for evaluation in Sect. 3. For example, in [25] the face image size is restricted to be at least 86×86 pixels, whilst a significant number of faces we use are of lower resolution.

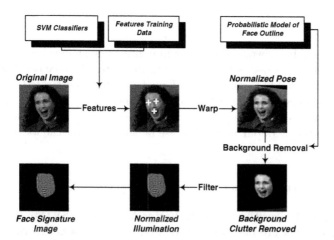

Fig. 3. Face representation: Each step in the proposed pre-processing cascade produces a result invariant to a specific extrinsic imaging factor. The result is a 'signature' image of a person's face

Table 1. Method overview: A summary of the main steps of the proposed algorithm. A novel, 'input' image **I** is first pre-processed to produce a signature image **S$_f$**, which is then compared with signatures of each 'reference' image **S$_r$**. The intermediate results of pre-processing are also shown in Fig. 3

Input:	novel face image **I**,
	training signature image **S$_r$**.
Output:	distance $d(\mathbf{I}, \mathbf{S}_r)$.

1: **Facial feature localization**
   $\{\mathbf{x}_i\} = \text{features}(\mathbf{I})$
2: **Pose effects: registration by affine warping**
   $\mathbf{I}_R = \text{affine\_warp}\,(\mathbf{I}, \{\mathbf{x}_i\})$
3: **Background clutter: face outline detection**
   $\mathbf{I}_F = \mathbf{I}_R. * \text{mask}(\mathbf{I}_R)$
4: **Illumination effects: band-pass filtering**
   $\mathbf{S} = \mathbf{I}_F * \mathbf{B}$
5: **Pose effects: registration refinement**
   $\mathbf{S}_f = \text{warp\_using\_appearance}(\mathbf{I}_F, \mathbf{S}_r)$
6: **Occlusion effects: robust distance measure**
   $d(\mathbf{I}, \mathbf{S}_r) = \text{distance}(\mathbf{S_r}, \mathbf{S_f})$

Everingham and Zisserman [94] consider AFR in situation comedies. However, rather than using facial feature detection, a quasi-3D model of the head is used to correct for varying pose. Temporal information via shot tracking is exploited for enriching the training corpus. In contrast, we do not use any temporal information, and the use of local features (Sect. 2.1) allows us to compare two face images in spite of partial occlusions (Sect. 2.5).

2 Method Details

In the proposed framework, the first step in processing a face image is the normalization of the subject's pose – *registration*. After the face detection stage, faces are only roughly localized and aligned – more sophisticated registration methods are needed to correct for the appearance effects of varying pose. One way of doing this is to "lock onto" characteristic facial points and warp images to align them. In our method, these facial points are the locations of the mouth and the eyes.

2.1 Facial Feature Detection

In the proposed algorithm Support Vector Machines[3] (SVMs) [40, 287] are used for facial feature detection. A related approach was described in [25];

---

[3] We used the LibSVM implementation freely available at
http://www.csie.ntu.edu.tw/~cjlin/libsvm/

alternative methods include pictorial structures [99], shape+appearance cascaded classifiers [105] and the method of Cristinacce et al. [72].

We represent each facial feature, i.e. the image patch surrounding it, by a feature vector. An SVM with a set of parameters (kernel type, its bandwidth and a regularization constant) is then trained on a part of the training data and its performance iteratively optimized on the remainder. The final detector is evaluated by a one-time run on unseen data.

2.1.1 Training

For training we use manually localized facial features in a set of 300 randomly chosen faces from the feature-length film "Groundhog day" and the situation comedy "Fawlty Towers". Examples are extracted by taking rectangular image patches centred at feature locations (see Figs. 4 and 5). We represent each patch $\mathbf{I} \in \mathbb{R}^{H \times W}$ with a feature vector $\mathbf{v} \in \mathbb{R}^{2H \times W}$ containing appearance and gradient information (we used $H = 17$ and $W = 21$ for a face image of the size 81×81 - the units being pixels):

$$v_A(Wy + x) = I(x, y) \tag{1}$$

$$v_G(Wy + x) = |\nabla I(x, y)| \tag{2}$$

$$\mathbf{v} = \begin{bmatrix} \mathbf{v}_A \\ \mathbf{v}_G \end{bmatrix} \tag{3}$$

Local Information

In the proposed method, implicit local information is included for increased robustness. This is done by complementing the image appearance vector $\mathbf{v}_A$ with the greyscale intensity gradient vector $\mathbf{v}_G$, as in Eq. (3).

Synthetic Data

For robust classification, it is important that training data sets are representative of the whole spaces that are discriminated between. In uncontrolled

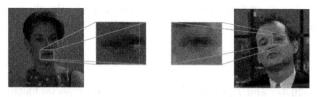

Fig. 4. Difficulties of facial feature detection: Without context, distinguishing features in low resolution and severe illumination conditions is a hard task even for a human. Shown are a mouth and an eye that although easily recognized within the context of the whole image, are very similar in isolation

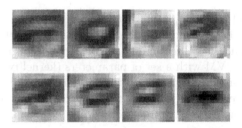

Fig. 5. SVM training data: A subset of the data (1800 features were used in total) used to train the SVM-based eye detector. Notice the low resolution and the importance of the surrounding image context for precise localization (see Fig. 4)

imaging conditions, the appearance of facial features exhibits a lot of variation, requiring an appropriately large training corpus. This makes the approach with manual feature extraction impractical. In our method, a large portion of training data (1500 out of 1800 training examples) was synthetically generated. Seeing that the surface of the face is smooth and roughly fronto-parallel, its 3D motion produces locally affine-like effects in the image plane. Therefore, we synthesize training examples by applying random affine perturbations to the manually detected set (for similar approaches to generalization from a small amount of training data see [11, 223, 316]).

2.1.2 SVM-Based Feature Detector

SVMs only provide classification decision for individual feature vectors, but no associated probabilistic information. Therefore, performing classification on all image patches produces as a result a binary image (a feature is either present or not in a particular location) from which only a single feature location is to be selected.

Our method is based on the observation that due to the robustness to noise of SVMs, the binary image output consists of connected components of positive classifications (we will refer to these as *clusters*), see Fig. 6. We use a prior on feature locations to focus on the cluster of interest. Priors corresponding to the three features are assumed to be independent and Gaussian (2D, with full covariance matrices) and are learnt from the training corpus of 300 manually localized features described in Sect. 2.1.1. We then consider the total 'evidence' for a feature within each cluster:

$$e(\mathcal{S}) = \int_{\mathbf{x} \in \mathcal{S}} P(\mathbf{x}) d\mathbf{x} \tag{4}$$

where $\mathcal{S}$ is a cluster and $P(\mathbf{x})$ the Gaussian prior on the facial feature location. An unbiased feature location estimate with $\sigma \approx 1.5$ pixels was obtained by choosing the mean of the cluster with largest evidence as the final feature location. Intermediate results of the method are shown in Fig. 6, while Fig. 7 shows examples of detected features.

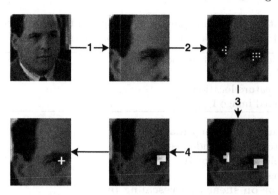

Fig. 6. Efficient SVM-based eye detection: 1: Prior on feature location restricts the search region. 2: Only ~ 25% of the locations are initially classified. 3: Morphological dilation is used to approximate the dense classification result from a sparse output. 4: The largest prior-weighted cluster is chosen as containing the feature of interest

Fig. 7. Automatic detection of facial features: High accuracy is achieved in spite of wide variation in facial expression, pose, illumination and the presence of facial wear (e.g. glasses and makeup)

2.2 Registration

In the proposed method dense point correspondences are implicitly or explicitly used in several stages: for background clutter removal, partial occlusion detection and signature image comparison (Sect. 2.3–2.5). To this end, images of faces are affine warped to have salient facial features aligned with their mean, canonical locations. The six transformation parameters are uniquely determined from three pairs of point correspondences – between detected facial features (the eyes and the mouth) and this canonical frame. In contrast to global appearance-based methods (e.g. [30, 90]) our approach is more robust to partial occlusion. It is summarized in Table 2 with typical results shown in Fig. 8.

Table 2. Registration: A summary of the proposed facial feature-based registration of faces and removal of face detector false positives

Input: canonical facial feature locations $\mathbf{x}_{can}$,
face image $\mathbf{I}$,
facial feature locations $\mathbf{x}_{in}$.
Output: registered image $\mathbf{I}_{reg}$.

1: **Estimate the affine warp matrix**
   $\mathbf{A} = \text{affine\_from\_correspondences}(\mathbf{x}_{can}, \mathbf{x}_{in})$
2: **Compute eigenvalues of $\mathbf{A}$**
   $\{\lambda_1, \lambda_2\} = \text{eig}(\mathbf{A})$
3: **Impose prior on shear and rescaling by $\mathbf{A}$**
   if $(|\mathbf{A}| \in [0.9, 1.1] \wedge \lambda_1/\lambda_2 \in [0.6, 1.3])$ **then**
4:      **Warp the image** $\mathbf{I}_{reg} = \text{affine\_warp}(\mathbf{I}; \mathbf{A})$
   else
5:      **Face detector false +ve**
   endif

Fig. 8. Effects of affine registration: Original (upper raw) and corresponding registered images (lower raw). The eyes and the mouth in all registered images are at the same, canonical locations. Registration transformations are significant

2.3 Background Removal

The bounding box of a face, supplied by the face detector, typically contains significant background clutter and affine registration boundary artefacts, see Fig. 8. To realize a reliable comparison of two faces, segmentation to foreground (i.e. face) and background regions has to be performed. We show that the face outline can be robustly detected by combining a prior on the face shape, learnt offline, and a set of measurements of intensity discontinuity in an image of a face. The proposed method requires only grey level information, performing equally well for colour and greyscale input, unlike previous approaches which typically use skin colour for segmentation (e.g. [12]).

In detecting the face outline, we only consider points confined to a discrete mesh corresponding to angles equally spaced at $\Delta\alpha$ and radii at Δr, see Fig. 9(a); in our implementation we use $\Delta\alpha = 2\pi/100$ and $\Delta r = 1$. At each

(a) 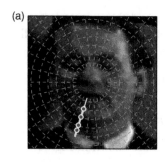 (b)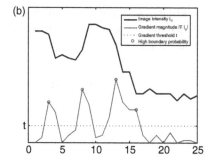

Fig. 9. Markov chain observations: (a) A discrete mesh in radial coordinates (only 10% of the points are shown for clarity) to which the boundary is confined. Also shown is a single measurement of image intensity in the radial direction and the detected high probability points. The plot of image intensity along this direction is shown in (b) along with the gradient magnitude used to select the high probability locations

mesh point we measure the image intensity gradient in the radial direction – if its magnitude is locally maximal and greater than a threshold t, we assign it a constant high-probability and a constant low probability otherwise, see Fig. 9(a,b). Let $\mathbf{m}_i$ be a vector of probabilities corresponding to discrete radius values at angle $\alpha_i = i\Delta\alpha$, and r_i the boundary location at the same angle. We seek the maximum *a posteriori* estimate of the boundary radii:

$$\{r_i\} = \arg\max_{\{r_i\}} P(r_1, .., r_N | \mathbf{m}_1, .., \mathbf{m}_N) = \tag{5}$$

$$\arg\max_{\{r_i\}} P(\mathbf{m}_1, .., \mathbf{m}_N | r_1, .., r_N) P(r_1, .., r_N), \text{ where } N = 2\pi/\Delta\alpha. \tag{6}$$

We make the Naïve Bayes assumption for the first term in equation Eq. (5), whereas, exploiting the observation that surfaces of faces are mostly smooth, for the second term we assume to be a first-order Markov chain. Formally:

$$P(\mathbf{m}_1, .., \mathbf{m}_N | r_1, .., r_N) = \prod_{i=1}^{N} P(\mathbf{m}_i | r_i) = \prod_{i=1}^{N} m_i(r_j) \tag{7}$$

$$P(r_1, .., r_N) = P(r_1) \prod_{i=2}^{N} P(r_i | r_{i-1}) \tag{8}$$

In our method model parameters (priors and likelihoods) are learnt from 500 manually delineated face outlines. The application of the model by maximizing expression in (5) is efficiently realized using dynamic programming i.e. the well-known Viterbi algorithm [119].

Feathering

The described method of segmentation of face images to foreground and background produces as a result a binary mask image $\mathbf{M}$. As well as masking the corresponding registered face image $\mathbf{I}_R$ (see Fig. 10), we smoothly suppress image information around the boundary to achieve robustness to small errors in its localization. This is often referred to as *feathering*:

$$\mathbf{M}_F = \mathbf{M} * \exp -\left(\frac{r(x,y)}{4}\right)^2 \tag{9}$$

$$I_F(x,y) = I_R(x,y)M_F(x,y) \tag{10}$$

Examples of segmented and feathered faces are shown in Fig. 11.

2.4 Compensating for Changes in Illumination

The last step in processing of a face image to produce its signature is the removal of illumination effects. A crucial premise of our work is that the most significant modes of illumination changes are rather coarse – ambient light varies in intensity, while the dominant illumination source is either frontal, illuminating from the left, right, top or bottom (seldom). Noting that these produce mostly slowly varying, low spatial frequency variations [103], we normalize for their effects by band-pass filtering, see Fig. 3:

$$\mathbf{S} = \mathbf{I}_F * \mathbf{G}_{\sigma=0.5} - \mathbf{I}_F * \mathbf{G}_{\sigma=8} \tag{11}$$

This defines the signature image $\mathbf{S}$.

Fig. 10. Face outline detection and background removal: Original image, image with detected face outline, and the resulting image with the background masked

Fig. 11. Face image segmentation: Original images of detected and affine-registered faces and the result of the proposed segmentation algorithm. Subtle variations of the face outline caused by different poses and head shapes are handled with high precision

2.5 Comparing Signature Images

In Sect. 2.1–2.4 a cascade of transformations applied to face images was described, producing a signature image insensitive to illumination, pose and background clutter. We now show how the accuracy of facial feature alignment and the robustness to partial occlusion can be increased further when two signature images are compared.

2.5.1 Improving Registration

In the registration method proposed in Sect. 2.2, the optimal affine warp parameters were estimated from three point correspondences in 2D. Therefore, the 6 degrees of freedom of the affine transformation were uniquely determined, making the estimate sensitive to facial feature localization errors. To increase the accuracy of registration, we propose a dense, appearance-based affine correction to the already computed feature correspondence-based registration.

In our algorithm, the corresponding characteristic regions of two faces, see Fig. 12(a), are perturbed by small translations to find the optimal *residual shift* (i.e. that which gives the highest normalized cross-correlation score between the two overlapping regions). These new point correspondences now *overdetermine* the residual affine transformation (which we estimate in the least L_2 norm of the error sense) that is applied to the image. Some results are shown in Fig. 12.

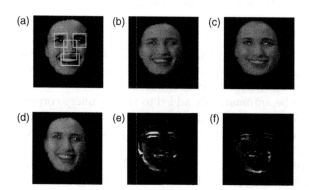

Fig. 12. Pose refinement: (a) Salient regions of the face used for appearance-based computation of the residual affine registration. (b)(c) Images aligned using feature correspondences alone. (d) The salient regions shown in (a) are used to refine the pose of (b) so that it is more closely aligned with (c). The residual rotation between (b) and (c) is removed. This correction can be seen clearly in the difference images: (e) is $|\mathbf{S}_c - \mathbf{S}_b|$, and (f) is $|\mathbf{S}_c - \mathbf{S}_d|$

2.5.2 Distance

Single Query Image

Given two signature images in precise correspondence (see above), $\mathbf{S}_1$ and $\mathbf{S}_2$, we compute the following distance between them:

$$d_S(\mathbf{S}_1, \mathbf{S}_2) = \sum_x \sum_y h(S_1(x,y) - S_2(x,y)) \tag{12}$$

where $h(\Delta S) = (\Delta S)^2$ if the probability of occlusion at (x,y) is low and a constant value k otherwise. This is effectively the L_2 norm with added outlier (e.g. occlusion) robustness, similar to [29]. We now describe how this threshold is determined.

Partial Occlusions

Occlusions of imaged faces in films are common. Whilst some research has addressed detecting and removing specific artefacts only, such as glasses [171], here we give an alternative non-parametric approach, and use a simple appearance-based statistical method for occlusion detection. Given that the error contribution at (x,y) is $\varepsilon = \Delta S(x,y)$, we detect occlusion if the probability $P_s(\varepsilon)$ that ε is due to inter- or intra- personal differences is less than 0.05. Pixels are classified as occluded or not on an independent basis. $P_s(\varepsilon)$ is learnt in a non-parametric fashion from a face corpus with no occlusion.

The proposed approach achieved a reduction of 33% in the expected within-class signature image distance, while the effect on between-class distances was found to be statistically insignificant.

Multiple Query Images

The distance introduced in Eq. (12) gives the confidence measure that two signature images correspond to the same person. Often, however, more than a single image of a person is available as a query: these may be supplied by the user or can be automatically added to the the query corpus as the highest ranking matches of a single image-based retrieval. In either case we want to be able to quantify the confidence that the person in the novel image is the same as in the query *set*.

Seeing that the processing stages described so far greatly normalize for the effects of changing pose, illumination and background clutter, the dominant mode of variation across a query corpus of signature images $\{\mathbf{S}_i\}$ can be expected to be due to facial expression. We assume that the corresponding manifold of expression is linear, making the problem that of point-to-subspace matching [29]. Given a novel signature image $\mathbf{S}_N$ we compute a robust distance:

$$d_G(\{\mathbf{S}_i\}, \mathbf{S}_N) = d_S(\mathbf{F}\mathbf{F}^T\mathbf{S}_N, \mathbf{S}_N) \tag{13}$$

where $\mathbf{F}$ is orthonormal basis matrix matrix corresponding to the linear subspace that explains 95% of energy of variation within the set $\{\mathbf{S}_i\}$.

3 Evaluation and Results

The proposed algorithm was evaluated on automatically detected faces from the situation comedy "Fawlty Towers" ("A touch of class" episode), and feature-length films "Groundhog Day" and "Pretty Woman"[4]. Detection was performed on every 10th frame, producing respectively 330, 5500, and 3000 detected faces (including incorrect detections). Face images (frame regions within bounding boxes determined by the face detector) were automatically resized to 80 × 80 pixels, see Fig. 15(a).

3.1 Evaluation Methodology

Empirical evaluation consisted of querying the algorithm with each image in turn (or image set for multiple query images) and ranking the data in order of similarity to it. Two ways of assessing the results were employed – using Receiver Operator Characteristics (ROC) and the *rank ordering score* ρ quantifying the goodness of a similarity-ranked ordering of data.

Rank Ordering Score

Given that data is recalled with a higher recall index corresponding to a lower confidence, the normalized sum of indexes corresponding to in-class faces is a meaningful measure of the recall accuracy. We call this the rank ordering score and compute it as follows:

$$\rho = 1 - \frac{S - m}{M} \qquad (14)$$

where S is the sum of indexes of retrieved in-class faces, and m and M, respectively, the minimal and maximal values S and $(S - m)$ can take. The score of $\rho = 1.0$ can be seen to correspond to orderings which correctly cluster all the data (all the in-class faces are recalled first), 0.0 to those that invert the classes (the in-class faces are recalled last), while 0.5 is the expected score of a random ordering. The *average normalized rank* [279] is equivalent to $1 - \rho$.

3.2 Results and Discussion

Typical Receiver Operator Characteristic curves obtained with the proposed method are shown in Fig. 13(a, b). These show that excellent results are obtained using as little as 1-2 query images, typically correctly recalling 92% of the faces of the query person with only 7% of false retrievals. As expected, more query images produced better retrieval accuracy, also illustrated in Fig. 13(e, f). Note that as the number of query images is increased, not only

---

[4] Available at http://www.robots.ox.ac.uk/~vgg/data/

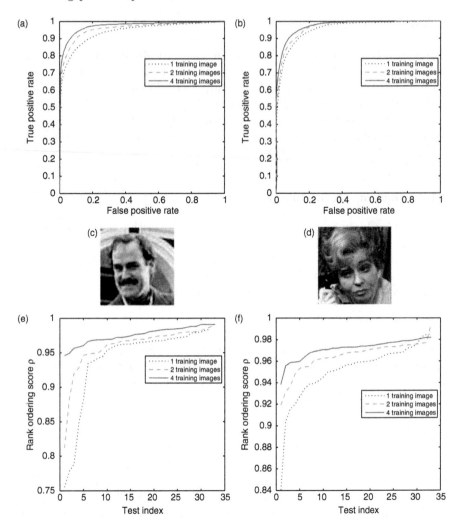

Fig. 13. Performance evaluation: (a, b) ROC curves for the retrieval of Basil (c) and Sybil (d) in "Fawlty Towers". The corresponding rank ordering scores across 35 retrievals are shown in (e) and (f), sorted for the ease of visualization

is the ranking better on average but also more robust, as demonstrated by a decreased standard deviation of rank order scores. This is very important in practice, as it implies that less care needs to be taken by the user in the choice of query images. For the case of multiple query images, we compared the proposed subspace-based matching with the k-nearest neighbours approach, which was found to consistently produce worse results. The improvement of recognition with each stage of the proposed algorithm is shown in Fig. 14.

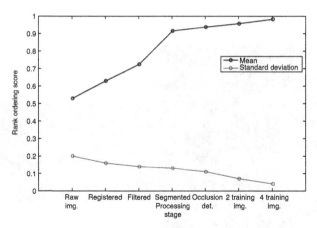

Fig. 14. Performance evaluation: The average rank ordering score of the baseline algorithm and its improvement as each of the proposed processing stages is added. The improvement is demonstrated both in the increase of the average score, and also in the decrease of its standard deviation averaged over different queries. Finally, note that the averages are brought down by few very difficult queries, which is illustrated well in Fig. 13 (e,f)

Example retrievals are shown in Figs. 15-17. Only a single incorrect face is retrieved in the first 50, and this is with a low matching confidence (i.e. ranked amongst the last in the retrieved set). Notice the robustness of our method to pose, expression, illumination and background clutter.

4 Summary and Conclusions

In this chapter we have introduced a content-based film-shot retrieval system driven by a novel face recognition algorithm. The proposed approach of systematically removing particular imaging distortions – pose, background clutter, illumination and partial occlusion has been demonstrated to consistently achieve high recall and precision rates on several well-known feature-length films and situation comedies.

The main research direction we intend to pursue in the future is the development of a flexible model for learning person-specific manifolds, for example due to facial expression changes. Another possible improvement to the method that we are considering is the extension of the current image-to-image and set-to-image matching to set-to-set comparisons. Previous work has shown set-to-set matching has inherent advantages in terms of identification accuracy [12, 290]. For video or films, sets of faces for each person can be gathered easily and automatically by tracking through shots, an idea that is explored in [295].

(a)

(b)

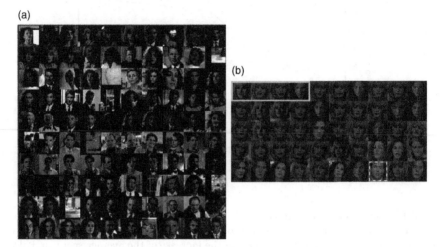

Fig. 15. Example retrieval: (a) The "Pretty Woman" data set – every 50th detected face is shown for compactness. Typical retrieval result is shown in (b) – query images are outlined by a solid line, the incorrectly retrieved face by a dashed line. The performance of our algorithm is very good in spite of the small number of query images used and the extremely difficult data set – this character frequently changes wigs, makeup and facial expressions

(a)

(b)

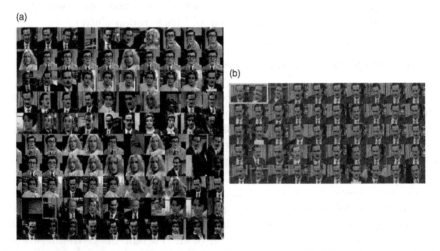

Fig. 16. Example retrieval: (a) The "Fawlty Towers" data set – every 30th detected face is shown for compactness. Typical retrieval result is shown in (b) – query images are outlined. There are no incorrectly retrieved faces in the top 50

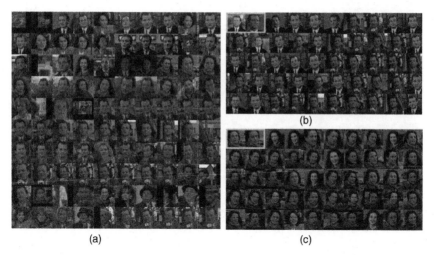

Fig. 17. Example retrieval: (a) The "Groundhog Day" data set – every 30th detected face is shown for compactness. Typical retrieval results are shown in (b) and (c) – query images are outlined. There are no incorrectly retrieved faces in the top 50

Acknowledgements

The authors would like to express their gratitude to to Mark Everingham for a number of helpful discussions and suggestions, and Krystian Mikolajczyk and Cordelia Schmid of INRIA Grenoble who supplied face detection code. Our thanks also go to Toshiba Corporation and Trinity College, Cambridge for their kind support of our research. Funding was provided by EC Project CogViSys.

Visual Audio: An Interactive Tool for Analyzing and Editing of Audio in the Spectrogram

C.G. van den Boogaart and R. Lienhart

Multimedia Computing Lab, University of Augsburg, 86159 Augsburg, Germany.
{boogaart,lienhart}@informatik.uni-augsburg.de

1 Introduction

Hearing, analyzing and evaluating sounds is possible for everyone. The reference-sensor for audio, the human ear, is of amazing capabilities and high quality. In contrast editing and synthesizing audio is an indirect and non-intuitive task, which needs great expertise. It is normally performed by experts using specialized tools for audio-effects such as a low-pass filter or a reverb. This situation is depicted in Fig. 1: A user can edit a given sound by sending it through an audio-effect (1). The input (2) and the output (3) are evaluated acoustically and sometimes but rarely also with a spectrogram (4,5). The audio-effects can only be controlled via some dedicated parameters (6) and therefore allow editing on a very abstract and crude level. In order to generate best results with this technique it is state of the art to record every sound separately on a different track in clean studio conditions. The effects can now be applied to each channel separately. More direct audio editing is desirable, but not yet possible.

The goal of Visual Audio is to lower these limitations by providing a means to directly and visually edit audio spectrograms, out of which high quality audio can be reproduced. Figure 2 shows the new approach: A user can edit the spectrogram of a sound directly. The result can be evaluated either visually or acoustically, resulting in a shorter closed loop for editing and evaluating. This has several advantages:

1. A spectrogram is a very good representation of an audio-signal. Often speech-experts are able to read text out of speech-spectrograms. In our approach, the spectrogram is used as a representation of both, the original and the recreated audio-signal, which both can be represented visually and acoustically. It therefore narrows the gap between hearing and editing audio.

2. Audio is transient. It is emitted by a source through a dynamic process, travels through the air and is received by the human ear. It cannot be

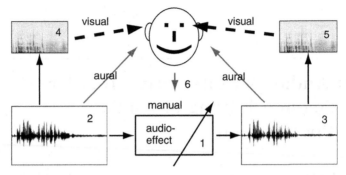

Fig. 1. Classical situation in audio editing: A sound is sent through an audio-effect (1). The input (2) and the output (3) are evaluated acoustically and sometimes visually (4,5). The audio-effects are controlled via some dedicated parameters (6)

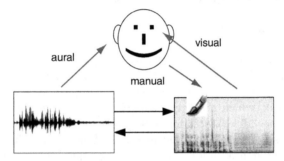

Fig. 2. Visual Audio: The spectrogram of a sound is edited directly. The result can be evaluated either visually or acoustically

frozen for investigation at a given point in time and a given frequency band. This limitation is overcome by representing the audio signal as a spectrogram. The spectrogram can be studied in detail and edited appropriately before transforming back into the transient audio domain.

Figure 3 gives an overview over the several stages of Visual Audio Editing. A time signal (1) is transformed (2) into one of a manifold of time-frequency representations (3). One representation is chosen (4) and edited (5). By inverse transformation (6) an edited time signal (7) is reproduced. By the appropriate choice of one of the manifold time-frequency representations, which refer to higher time or higher frequency resolution, it is possible to edit with high accuracy in time and frequency. If necessary, the process is repeated (8). Figure 4 shows a screenshot of our implementation of Visual Audio Editing.

Related Work: Time-frequency transformations are a well-known and intensively used tool in automatic processing of audio signals. Standard transformations are: wavelets [75, 266], DFT and FFT [244] and Wigner-Ville-Distribution [257]. The Gabor transformation [107] is closely related to the MLT (Modulated lapped transform) [218], which has many applications in audio processing. The time-frequency transformations are mainly used either

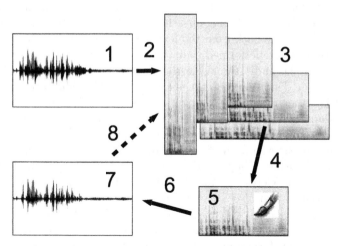

Fig. 3. Overview of Visual Audio Editing: a time signal (1) is transformed (2) into one of a manifold of time-frequency representations (3). One representation is chosen (4) and edited (5). By inverse transformation (6) an edited time signal (7) is recreated. If necessary, the process is repeated (8)

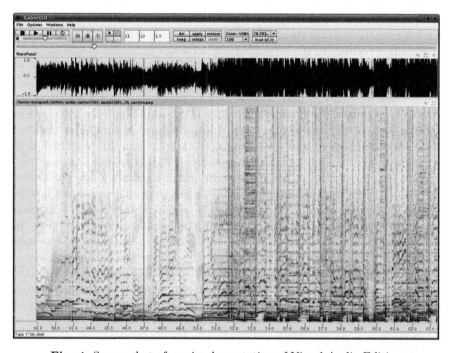

Fig. 4. Screen shot of our implementation of Visual Audio Editing

for calculating features or for compression and filtering. In the first case, the features are used for recognizing speech [149], speakers [263] or music pieces [123]. There is no way back from the features to the audio signal. An inverse transformation is not performed, but the features are used to derive higher semantics from the audio signal. In the second case, e.g. lossy speech compression [266] or denoising [86], an inverse transformation is performed and the signal is either intentionally or unintentionally edited in the short term frequency domain. However the editing is not based on any visualization and thus no visual manipulation concepts can be applied.

Nevertheless the approach of editing audio in its spectrogram has already been presented earlier. One is Audiosculpt from IRCAM[1] followed by Ceres, Ceres2 and last by Ceres3[2], which are designed for musicians to create experimental sounds and also for education. They work as FFT/IFFT analysis/resynthesis packages, which allow editing the short time Fourier transformation spectrogram of an audio signal. The user has to choose several parameters for transformation and reconstruction, e.g. the window shape itself, but is restricted to very few fixed window lengths. Another approach is reported by Horn [145]. Is is based on auditory spectrograms, which model the spectral-transformation of the ear and is dedicated to speech, i.e. only for a small bandwidth. The spectrogram is first abstracted to the so called part-tone-time-pattern and then edited and reconstructed.

There are two main differences compared to Visual Audio Editing as we present it here: First we use the Gabor transformation with the Gaussian window (see [107]). This transformation is optimal in terms of time-frequency resolution according to the Heisenberg uncertainty principle as well as in reconstruction quality. The uncertainty principle allows choosing the ratio of time to frequency-resolution. Therefore the user is allowed to choose this himself continuously and as only transformation parameter. In the following we refer to it as resolution zooming operation. As sound images are more or less complex structures, we secondly introduce techniques for smart user assisted editing by the usage of template sounds, so called audio objects, which help to structure and handle the sound image.

This chapter is organized as follows: In Sect. 2 we explain how the Gabor transformation is used to create an image out of the audio signal and how to recreate the audio signal from the image. A single parameter, the resolution zooming factor, is used in order to adapt the time-frequency resolution to the task at hand. In Sect. 3 the manual editing of the image by the user is described. In Sect. 4 we discuss and give examples on smart user assisted editing. This is enabled by the use of audio objects, i.e. templates, which allow interacting with the image, preserving its complex structure and therefore preserving the high audio quality. Sect. 5 concludes with a short summary of the main aspects.

---

[1] http://www.ircam.fr

[2] http://music.columbia.edu/ stanko/About_Ceres3.html

2 From Audio to Visual Audio and Back Again

An audio signal is in general given in the 1D time-domain. In order to edit it visually, a 2D representation is necessary, which gives the user descriptive information about the signal and out of which the original or edited signal can be reconstructed. In this section we therefore discuss how to convert an audio signal into the image domain and how to recreate audio from that image domain.

2.1 Time-Frequency Transformation: Gabor Transformation

As image domain we use a spectrogram. The spectrogram of an audio signal is called imaged sound. It shows the magnitude of a time-frequency transformation of the audio signal. Time-frequency transformations reveal local properties of a signal and allow to recreate the signal under certain conditions. The kind of revealed properties, however, depend strongly on the window and the window length. We use the Gabor transformation with a Gaussian window and apply multiwindow techniques to find the best matching window length for a given task (see e.g. [342]).

Fundamentals of the Gabor transformation: The Gabor transformation was introduced by Dennis Gabor [107] and has gained much attention in the near past (see e.g. [97] and [98]). In conjunction with the Gaussian window, which is not the only possible choice, the Gabor transformation has perfect time-frequency localization properties. It splits up a time function $x(t)$ in its time-frequency representation $X(t, f)$ and is defined as follows: From a single prototype or windowing function $g(t)$, which is localized in time and frequency, a Gabor system $g_{na,mb}(t)$ is derived by time shift a and frequency shift b (see [313]):

$$g_{na,mb}(t) = e^{2\pi jmbt}g(t - na), \; n, m \in \mathbb{Z}, \; a, b \in \mathbb{R}, \tag{1}$$

a and b are called the lattice constants. The Gabor system covers the whole time-frequency plane. The Gabor transformation is then expressed as follows:

$$c_{nm} = X(na, mb) = \int_{-\infty}^{+\infty} x(t)g_{na,mb}^*(t)\, dt. \tag{2}$$

c_{nm} are called the Gabor coefficients of $x(t)$. The inverse transformation or Gabor expansion is calculated with the window function $\gamma(t)$, which is called dual window of $g(t)$. With the Gabor system $\gamma_{na,mb}(t)$ defined analogously to equation (1), the Gabor expansion is defined as follows (see [23]):

$$x(t) = \frac{1}{L} \sum_{n=-\infty}^{\infty} \sum_{m=-\infty}^{\infty} c_{nm}\gamma_{na,mb}(t) \tag{3}$$

with $L = \sum_{k=-\infty}^{\infty} |\gamma(ka)|^2$. The theory of the Gabor transformation leads to the result, that the window functions $g(t)$ and $\gamma(t)$ and the lattice constants a and b must fulfill certain requirements in order to assure the invertibility (see [98]). The appropriate choices are discussed in the following two paragraphs. For use in digital signal processing formulas (2) and (3) have to be discretized, using sums instead of integrals and sums of finite length, which is discussed in the last paragraph of this section.

The Gaussian window: We start with the discussion of the Gaussian window function, which is given as:

$$g(t) = \frac{1}{\sqrt{2\pi\sigma_t^2}} e^{-\frac{1}{2}\frac{t^2}{\sigma_t^2}} \tag{4}$$

and has the following advantageous properties (see e.g. [360]):

- Minimal extent in the time-frequency plane according to the Heisenberg uncertainty principle.
- Localized shape, i.e. only one local and global maximum with strict decay in time and frequency direction.

The uncertainty principle of Heisenberg says that the product of temporal and frequency extent of a window function has a total lower limit. If the extents are defined in terms of standard deviations of the window function and of its Fourier transformation respectively, it can be expressed with the following inequality (see [266]):

$$\sigma_t \sigma_f \geq \frac{1}{4\pi} . \tag{5}$$

The "=" is only reached for the Gaussian window function (see e.g. [321]), which therefore has the minimal possible extent in the time-frequency plane. Its Fourier transformation has the same Gaussian shape as the time function itself:

$$G(f) = \frac{1}{\sqrt{2\pi\sigma_f^2}} e^{-\frac{1}{2}\frac{f^2}{\sigma_f^2}}, \text{ with } \sigma_f = \frac{1}{4\pi\sigma_t}. \tag{6}$$

The dual window $\gamma(t)$ should also be localized in time and frequency to preserve the local influence of the Gabor coefficients on the result of the inverse transformation. Also this is best achieved by the Gaussian as dual window-function. To ensure perfect reconstruction some restrictions are imposed on the choice of lattice constants a and b as discussed in the following paragraph.

Choice of lattice constants and oversampling factor: In his original paper [107] Dennis Gabor suggested to choose $ab = 1$ which is called critical sampling. This choice has implicit influence on the shape of the dual window. In fact with the Balian-Low theorem it can be shown that in this case the dual window extends to infinity and is not localized at all (see [98]). The solution is to choose $ab < 1$, which is referred to as the oversampled case. It leads to better localized dual windows and numerically stable analysis and synthesis.

We therefore have to determine an appropriate oversampling factor for $ab < 1$. In the literature normally the cases of rational oversampling ($ab = \frac{p}{q}$, $p, q \in \mathbb{N}$ and $p < q$) and integer oversampling ($ab = \frac{1}{q}$, $q \in \mathbb{N}$) are discussed. Bastiaans [23] proposes taking $ab = \frac{1}{3}$ for which the ideal dual window of the Gaussian is very similar to a Gaussian window. He mentions that for increasing values of q the resemblance of the Gaussian window and its dual window further increases. In simple empirical hearing tests we found that an oversampling factor starting at $ab = \frac{1}{5}$ avoids hearable differences between an original and a reconstructed sound in case of using the same Gaussian window for analysis and synthesis. This holds for speech and for full bandwidth music. An oversampling factor of $q = 5$ is therefore the necessary and sufficient accuracy for high quality audio processing.

In the following, if necessary we name a, b for the critical sampled case a_{crit}, b_{crit} and for the oversampled case a_{over}, b_{over}. The resulting lattice and how it covers the time-frequency plain is illustrated in Fig. 5. The gray shaded circles indicate the extent of a single Gaussian window in the time-frequency plain expressed in its standard deviations σ_t and σ_f. To ensure the same overlapping of the Gaussians in time and frequency direction, we therefore have to set:

$$\frac{\sigma_t}{\sigma_f} = \frac{a_{crit}}{b_{crit}} = \frac{a_{over}}{b_{over}}. \tag{7}$$

With $a_{over}b_{over} = \frac{1}{q}$ this holds for:

$$a_{over} = \frac{a_{crit}}{\sqrt{q}}, b_{over} = \frac{b_{crit}}{\sqrt{q}}. \tag{8}$$

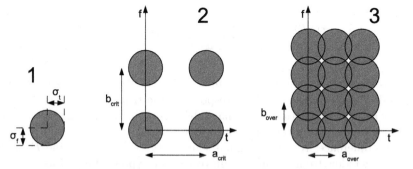

Fig. 5. Coverage of the time-frequency plain: The gray shaded circles indicate the extent of a (1) single Gaussian window expressed in its standard deviations σ_t and σ_f, (2) the critically sampled case and (3) the oversampled case

With formula (6) and formula (7) we can solve:

$$\sigma_t = \frac{1}{\sqrt{4\pi}}\sqrt{\frac{a}{b}}, \sigma_f = \frac{1}{\sqrt{4\pi}}\sqrt{\frac{b}{a}}. \tag{9}$$

As in the following we will always choose b and determine the other values, we continue with $ab = \frac{1}{q}$:

$$\sigma_t = \frac{1}{\sqrt{4\pi q}}\frac{1}{b}, \sigma_f = \sqrt{\frac{q}{4\pi}}b. \tag{10}$$

With an oversampling factor of $q = 5$ and these formulas, it is feasible to take the Gaussian window $g(t)$ as its own dual window $\gamma(t) = g(t)$. One still has the freedom to choose either a_{crit} or b_{crit}, i.e. to choose an appropriate window length for the current task.

Discretization and Truncation: The formulas (2) and (3) are for a continuous representation of $x(t)$ and calculate sums and integrals over infinity. Therefore they have to be discretized and the sums and integrals have to be truncated in order to be implemented. This corresponds to bandlimiting and sampling $x(t)$ and to truncating $g(t)$. $x(t)$ is sampled with the sampling frequency f_s. With $T = 1/f_s$ and $k \in \mathbb{Z}$ $x(t)$ becomes $x(kT)$. To fulfill the sampling theorem, the bandwidth f_B of $x(kT)$ has to fulfill $f_B \leq f_s/2$ (see e.g. [244])[3]. We define $M \in \mathbb{N}$ as the number of frequency bands with $M = \lceil \frac{1}{2}\frac{f_s}{b_{over}} \rceil$. With t_{cut} half the window length, we define $N \in \mathbb{N}$ with $N = t_{cut}f_s = \frac{t_{cut}}{T}$ as half the window length in samples. The Gabor transformation can then be implemented as:

$$c_{nm} = X(na, mb) = \frac{1}{L}\sum_{k=-N}^{N-1} x(kT)g^*_{na,mb}(kT) \tag{11}$$

with its inverse:

$$x(kT) = \frac{1}{L}\sum_{n=\lfloor \frac{kT-t_{cut}}{a} \rfloor}^{\lceil \frac{kT+t_{cut}}{a} \rceil}\sum_{m=0}^{M-1} c_{nm}g_{na,mb}(kT) \tag{12}$$

where $L = \sqrt{q\sum_{k=-N}^{N-1}|g(ka)|^2}$ and $m \in [0, M-1]$. We still have to determine t_{cut} and N respectively, which correspond to the truncation of the Gaussian window as already mentioned. We have chosen empirical listening tests to find an appropriate truncation. The goal was to get a reproduced sound with no hearable difference from the original sound. We define the decline D of the Gaussian window from the maximum to the cut expressed in dB as:

$$D = 20log\frac{g(0)}{g(t_{cut})}dB. \tag{13}$$

---

[3] For hi-fi audio typical values are $f_s = 44.1kHz$ (CD) or higher and $f_B = 20kHz$.

For a given D we get with (4):

$$t_{cut} = \sqrt{2ln\left(10^{\frac{D}{20}}\right)}\sigma_t. \tag{14}$$

Values of $D \geq 30dB$ have shown to be completely sufficient for high quality audio.

Remark: Storing the Gabor coefficients is very efficient independent of the time-frequency resolution. A discretized time signal of duration t_{dur} needs $N^{\mathbb{R}} = t_{dur}f_s$ real sample values. The Gabor coefficients need

$$N_{Gabor}^{\mathbb{C}} = \frac{t_{dur}}{a}\frac{f_B}{b} = t_{dur}f_B q \tag{15}$$

complex or

$$N_{Gabor}^{\mathbb{R}} = 2 \cdot N_{Gabor}^{\mathbb{C}} = 2t_{dur}f_B q \tag{16}$$

real and imaginary values combined for storage. In summary $N_{Gabor}^{\mathbb{R}} = q \cdot N^{\mathbb{R}}$ values independent of the current time-frequency resolution are needed to store the Gabor transformation. This can be further enhanced by setting $f_B = 20kHz$ with $f_B < f_s/2$.

2.2 One Degree of Freedom: Resolution Zooming

The human ear has a time-frequency resolution, which closely reaches the physical limit expressed in the Heisenberg uncertainty principle. It furthermore adapts its current time-frequency resolution to the current content of the signal according to the Heisenberg uncertainty principle (see [32]). It is therefore advantageous also to adapt the resolution of the transformation to the current editing task. The resolution zooming feature is discussed in this section.

Heisenberg uncertainty principle: We already applied the Heisenberg uncertainty principle to the functions of the window and the dual window (see Sect. 2.1). It also holds for the function of the signal, which of course in general has a worse resolution than the theoretical limit.

As the Gabor transformation is a discretized version of the STFT, we can discuss the continuous case of the STFT. The STFT can be expressed as a convolution of signal $x(t)$ with $h^*(-t, f) = w^*(-t)e^{-j2\pi ft}$:

$$X(t,f) = x(t) * h^*(-t,f) = \int_{-\infty}^{+\infty} x(\tau)h^*(-(t-\tau),f)\,d\tau. \tag{17}$$

This is similar to adding the variances of two statistically independent random variables X and Y, which form a new random variable $Z = X + Y$. Their probability density functions are also convolved and the variance of Z is then given by (see [137]):

$$\sigma_Z^2 = \sigma_X^2 + \sigma_Y^2. \tag{18}$$

Therefore the time and frequency variances of a signal and a window are added in the form:

$$\sigma^2_{t_{transformation}} = \sigma^2_{t_{signal}} + \sigma^2_{t_{window}}, \tag{19}$$

$$\sigma^2_{f_{transformation}} = \sigma^2_{f_{signal}} + \sigma^2_{f_{window}}. \tag{20}$$

Consequently the achievable accuracy of editing a given signal is given by the superposition of the signal's and window's uncertainty. As time and frequency resolution are interconnected, one has to give up time resolution in order to improve the frequency resolution and vice versa.

Which window length to choose: So one has to choose the frequency shift b_{crit} resulting in a time shift $a_{crit} = \frac{1}{b_{crit}}$ or vice versa. This is equivalent to choosing an adapted window length. Different choices lead to different time-frequency representations of the same signal in the 3D-space with the axes t, f and e.g. b_{crit}. Although the same signal is represented, different characteristics of the signal are revealed in different layers with constant b_{crit}. The properties of this space can be clarified by the extremes of b_{crit}:

$b_{crit} \rightarrow \infty$: The window $g(t)$ becomes the Dirac impulse and the Gabor transformation becomes the time signal itself.

$b_{crit} = 0$: The window becomes $g(t) = const.$ losing its windowing properties and the Gabor transformation becomes the Fourier transformation.

In Visual Audio Editing it is therefore essential to calculate the Gabor transformation of an audio signal with different choices of b_{crit} and to perform the respective editing task in the layer $b_{crit} = const.$ which allows the best accuracy for the current task. This can be understood as zooming, which allows increasing the resolution of the representation of a signal either in time or in frequency, while the resolution of the other domain decreases.

From this discussion it also follows, that in contrast to images, the two axis time and frequency are not equivalent. A rotation of an image or of a region will lead to rather undesirable results and has to be omitted.

Example: Figure 6 a) shows a typical spectrogram of a speech signal. The spectrogram is calculated with $b_{crit} = 64Hz$, which results with $b_{over} = 28.6Hz$ in roughly 699 frequency bands in the range from $0Hz$ to $20kHz$. One property of the signal is hidden in this spectrogram: The recording was accidentally interfered by power line hum at $50Hz$ (common in Europe). The hum can be heard, if the sound is played at higher volume levels, but it cannot be seen in this spectrogram. Figure 6 b) shows a spectrogram of the same recording with a much higher frequency resolution of $b_{crit} = 5Hz$, i.e. $b_{over} = 2.24Hz$. In this spectrogram it is easy to distinguish the hum at $50Hz$ and his higher harmonics from the rest of the signal. Figure 6 c) shows the spectrogram with $b_{crit} = 64Hz$ again, but the image is "stretched" in frequency direction. The energy of the hum is distributed widely in the spectrogram and the higher harmonics are totally blurred and cannot be recognized at this frequency-resolution.

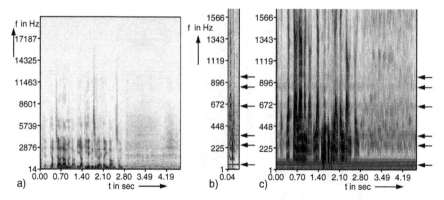

Fig. 6. Typical spectrograms for speech signal with an interfering power line hum. a): $b_{crit} = 64Hz$, b): $b_{crit} = 5Hz$, c): $b_{crit} = 64Hz$, "stretched" in frequency direction. For convenience b) and c) are cut above $1600Hz$. The interfering power line hum and higher harmonics can be recognized easily in the middle spectrogram. They are indicated by arrows on the right at $50Hz$, $250Hz$, $350Hz$, $650Hz$, $850Hz$ and $950Hz$. The energy of the hum is distributed widely in the right spectrogram and the higher harmonics are totally blurred

2.3 Fast Computation

The Gabor transformation as we use it here can be expressed as a special form of the DFT (Discretized Fourier Transform), if the window is understood as a part of the signal. At every multiple of the time shift a_{over} a DFT has to be computed. For hi-fi audio with high sampling frequency and some signal length, this leads to rather bad performance expressed in computation time. The DFT can be faster implemented as FFT (Fast Fourier Transform), if the the DFT length is a power of 2. As the DFT as then restricted to dedicated lengths, these forces the use of only some $b_{over} = \frac{f_s}{2M}$. They are moreover dependent on the sampling frequency f_s of the original signal. As for Visual Audio Editing b_{over} shall be chosen continuously the only possible solution up to now was to calculate a much more time consuming DFT.

We propose a solution to soften the hard restriction by the FFT window length, which allows to take advantage of the higher performance of the FFT. This introduces some computational overhead, which is generally acceptable if the overall execution time is still lower than that of a DFT, which is often the case.

For fast calculation of a DFT it is common practice to zero pad a given signal to a power of two boundary. By this action, the frequency spacing of the resulting values is also modified. We follow a very similar idea. By extending the window from a FFT boundary at $2M$ to a broader window at the FFT boundary $2\dot{M} = 2M \cdot 2^\eta$ ($\eta \in \mathbb{N}$), the density of frequency values is increased. Instead of $b_{over} = \frac{f_s}{2M}$ we get $\dot{b}_{over} = \frac{f_s}{2\dot{M}} = \frac{b_{over}}{2^\eta}$. Because of the frequency-shape of the window, which is not altered by altering the FFT

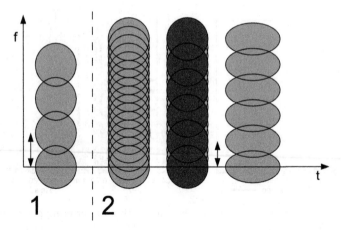

Fig. 7. This illustrates the downsampling. The vertical arrows mark in each case b_{over}. (1) shows a FFT with length $2M$ and the corresponding b_{over}. (2) shows a FFT with length $2\ddot{M} = 2M \cdot 2^2$. The blue circles mark a frequency vector generated by downsampling with factor 3. This alters the lattice constants. The resulting frequency vector with adapted σ_t and σ_f is shown

length, the frequency values can now be downsampled by a factor $\kappa \in \mathbb{N}$, $\kappa < 2^\eta$, in order to get a $\ddot{b}_{over} = \kappa\dot{b}_{over}$. Figure 7 explains the scenario.

Before performing the inverse transformation the skipped values have to be reconstructed. This can be done in the (complex) frequency signal very similar to the way it is done with (real) time signals: sample up by the factor κ and filter with the reconstruction filter. This filter is in our case the frequency representation of the Gaussian window (see formula (6)). As a convolution in the frequency domain is a multiplication in the time domain, this step can be done much more efficient after the inverse transformation in the time domain by multiplication with the time representation of the Gaussian window. Performance measurements and more details can be found in [33].

2.4 Visualization: Magnitude, Phase and Quantization

The transformation-data is represented as complex numbers with real and imaginary 32 bit float values. These numbers can not be visualized directly. Instead of complex numbers the magnitude values are used. They are then compressed and quantized to 8 bit unsigned integer values by calculating the square root of the values and scaling them (per image), to fit in 8 bit, while using the complete number range.

The processing tasks are performed on the transformation data, while the visualization is constantly updated. If an processing tasks is performed on the magnitude values only the phases are stored unchanged for the inverse transformation.

3 Manual Audio Brushing

We now describe, how imaged sounds can be edited manually. An obvious approach is to edit them like bitmaps – a well understood paradigm as documented by standard software packages such as Adobe Photoshop^{TM}. These techniques will allow us to perform tasks, which are either very complicated or even impossible with classical filtering techniques. Bitmap editing operations serve as a basis for developing and understanding of content based audio manipulation techniques. Our tool also allows to listen to selected regions of an imaged sound. This can be used to provide an instant feedback for the performed sound manipulations and thus serves as a perfect evaluation tool for the achieved sound quality.

3.1 Time-Frequency Resolution

The first task in editing is to find the right time-frequency resolution. The right resolution allows to select a given sound very accurately. This will be illustrated with three prototypical sounds. Figure 8 shows the imaged sound of music with three instruments: guitar, keyboard and drums. The time-frequency resolution is set to $b_{crit} = 52.41Hz$. The sound of a cymbal is marked with a rectangle. Because of the chosen time-frequency resolution the cymbal-sound can be found very compact in the spectrogram. For sounds with other characteristics different time-frequency resolutions have to be chosen.

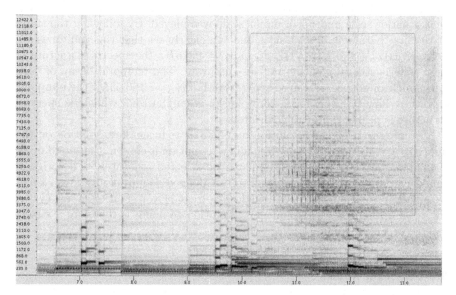

Fig. 8. Music with three instruments: guitar, keyboard and drums. Time-frequency resolution: $b_{crit} = 52.41Hz$. The sound of a cymbal is marked with a rectangle

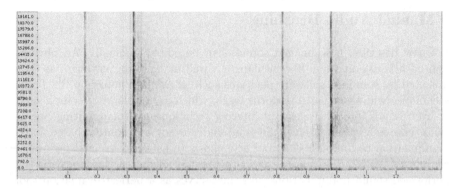

Fig. 9. Clicks of a ball-pen. Time-frequency resolution: $b_{crit} = 196.53Hz$. This sound has mainly transient components with very temporal characteristics

Figure 9 shows the imaged sound of clicks of a ball-pen with a time-frequency resolution set to $b_{crit} = 196.53Hz$, i.e. with a higher time resolution. This sound has mainly transient components and is very localized in time as can be seen clearly in the spectrogram.

A third example illustrates, that changing the time-frequency resolution not only changes the visualization but also more clearly reveals or hides important information. See Fig. 10, which shows the sound of a piano playing a C-major scale, each note separately. The imaged sound is represented in three different time-frequency resolutions: $b_{crit} = 10.65Hz$, $b_{crit} = 49.13Hz$ and $b_{crit} = 196.53Hz$. For $b_{crit} = 10.65Hz$ it is easy to identify the fundamental frequency and the higher harmonics of each note. It is even possible to verify, that a major scale and not a minor scale was played. By following the stairs of the first harmonic of each note one can clearly see that the semitones are between III, IV and VII, VIII. For $b_{crit} = 49.13Hz$ the spectral structure of each tone is still to identify, but less accurate. The temporal decay of each harmonic can now be perceived separately. For $b_{crit} = 196.53Hz$ the temporal structure, how fast the notes were played, is emphasized, while the spectral structure is nearly completely smeared.

Zooming to a higher resolution on one axis reduces the resolution on the opposite axis. If the right time-frequency resolution is chosen, the sound qualities of interest are separated and can be selected separately. The original sound can still be reconstructed from an imaged sound at any given time-frequency resolution. The combined time-frequency resolution is always at the total optimum. Time-frequency resolution zooming is therefore a strong feature of Visual Audio.

3.2 Selection Masks

Once you have selected the time-frequency resolution, a mask defining the sound pieces you want to edit must be constructed. In order to pick up different

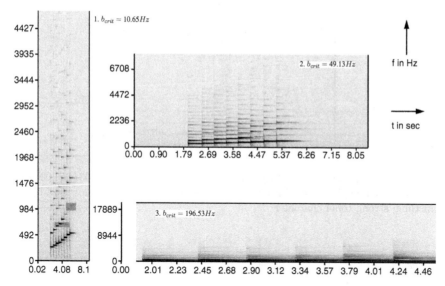

Fig. 10. Sound of a piano playing a C-major scale, each note separately. Time-frequency resolutions: 1. $b_{crit} = 10.65Hz$, 2. $b_{crit} = 49.13Hz$ and 3. $b_{crit} = 196.53Hz$

kinds of sounds, masks are necessary, which represent common structures of sounds. The better the mask matches a sound, the easier it is to select.

This already works with simple masks, because of two reasons: Firstly, similar physical generation mechanisms of different sounds of the same class have roughly the same shape. Secondly, the ear is robust to small degenerations in sound quality due to experience (recall the bad sound quality of a telephone compared to original speech) and due to masking effects (see [379]) near the edge of a selecting mask.

We review some sensible selecting masks with corresponding sound examples.

- rectangle masks
- comb masks
- polygon masks
- combination masks

Rectangle mask: Figure 8 already contained an example of a rectangle mask. In this case the sound of a cymbal is such localized in the spectrogram ($b_{crit} = 52.41Hz$), that it can easily be isolated by a rectangle. This can be verified by listening to the outer or the inner part of the rectangle only.

The rectangle mask furthermore exists in two extreme shapes. One is useful, in order to select sounds with very temporal characteristics, the other is useful for sounds with strong tonal character. Figure 11 shows left an example of a click of a ball-pen and right an example of a whistling sound.

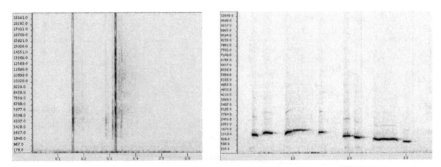

Fig. 11. Rectangle mask in the two extreme shapes. Left ($b_{crit} = 196.53Hz$): Click of a ball-pen. Rectangle mask with temporal characteristics. Right ($b_{crit} = 65.51Hz$): whistling, strong tonal character

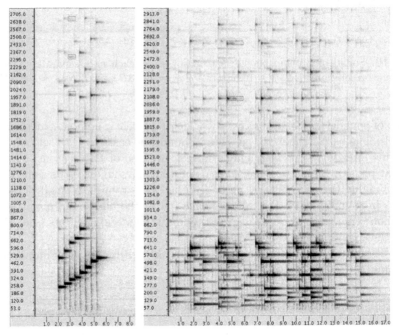

Fig. 12. Sound of a piano, $b_{crit} = 11.46Hz$. Left: C-major scale. Right: Short polyphonic piece

Comb masks: In contrast to the whistling sound, most tonal sounds not only incorporate the fundamental frequency, but also higher harmonics. This leads us to the comb masks. Figure 12 ($b_{crit} = 11.46Hz$) shows again the sound of a piano playing a C-major scale, each note separately (left). The mask useful in this case is called comb mask. As in this case a sound with a prominent pitch always generates a regular structure of higher harmonics which in its regularity is similar to a comb. A comb mask is defined by the following parameters: a function which describes the developing of the frequency of the highest harmonic, the number of harmonics and the bandwidth of every single harmonic. Figure 12 shows the spectrogram of a short piano piece ($b_{crit} = 11.46Hz$), which is of course polyphonic (right). The comb structure of each single note is nevertheless preserved.

Polygon masks: The last mask we want to explain simply uses polygons. See Fig. 13 for an example. It shows a speech signal with background noise ($b_{crit} = 78.61Hz$). Polygons allow to select such complex regions, while requiring more elaborateness of the user. They can be used e.g. to select structured and desired sounds between of surrounding broadband noise.

Combination masks: To match for complex sounds it is possible to build up complex masks out of repeated simple masks of the same type or simple masks of different types.

3.3 Interaction

Once a sound of interest is selected with an appropriate mask, it is possible to edit the imaged sound. Their are several useful possibilities.

Amplifying, stamping: The simplest editing is to multiply the magnitude values with a certain factor A. For $A = 0$ the sound is erased or stamped out, for $0 < A < 1$ the sound is damped in its level, for $A = 1$ it is unchanged and for $A > 1$ the sound is amplified.

Equalization: If a factor $A(f, t)$ as function of time and frequency is used, the result is similar to a very complex equalization process.

Cut, Copy&Paste: It is also possible to move the sound around in the spectrogram. If only the time position is changed, the same sound is just moved or copied to an other time instance.

Changing the pitch: If a sound is moved not only in time, but also in frequency, this changes the pitch of the sound too.

Changing the shape: If the sound has harmonic components, it is necessary in this case, to stretch or compress the sound in frequency direction, to preserve the harmonic structure. Some more details of this issue are discussed in the next section under the term "image-transformations"

Evaluating: A very helpful mechanism in Visual Audio is the playback of selected regions of an imaged sound. In practice it is not trivial to choose the correct shape for a mask for a given desired sound operation. The accuracy of the shape can however easily be evaluated by simply listening to the parts inside and the parts outside the mask respectively. By the presence or absence of a sound quality in these two parts of the sound, it can be clearly distinguished, whether the shape of the mask has to be tuned further or whether it is already correct.

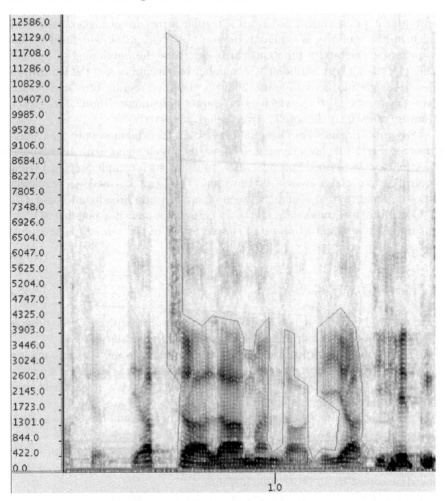

Fig. 13. Imaged sound of speech with background noise, $b_{crit} = 78.61Hz$. The polygon selects parts of the speech signal, separating it from the surrounding broadband noise

4 Smart Audio Brushing

In this section we will discuss smart techniques for Visual Audio Editing, which are based on audio objects. This is a more flexible approach than the one presented in the preceding chapter, because audio objects can adapt to much more complex shapes, than simple masks.

Firstly we will discuss the features of audio objects. A sound recorded beforehand under defined conditions is used as an audio object i.e. as a template mask for editing the current audio track.

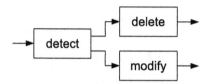

Fig. 14. Two-step editing process for audio objects

Secondly we illustrate the detection of known audio objects. Detections are described by the locations (i.e. positions and shapes) of the audio objects in the spectrograms. Detection has to be performed resilient to typical variations in which audio objects of interest can be experienced.

Last we explain deleting and modifying of detected audio objects. Audio object deletion is a special form of modifying audio objects. An example for a more general modification is editing the audio object in a different track before adding it to the original audio track again. The remaining signal content is at the same time left unchanged. Figure 14 summarizes these two possible two-step editing processes.

4.1 Audio Objects: Template Masks

Reproducible sounds, which are well-structured can be treated as visual objects in the spectrogram. In the spectrogram, they are characterized by a distinctive visual pattern – a pattern which looks similar even under typical variations such as frequency shifts, different play rates, and recordings from different microphones, different rooms, and playback devices. Examples could be individual notes played by an instrument, the sound of an closing car door or the rattling of a single key on a PC-keyboard. An example is also the click of a ball-pen, whose imaged sound we already discussed in Fig. 9 and Fig. 11. An other example is an individual note played by a piano. See for instance Fig. 15. If the structure of a sound is stable over long times, even longer portions of a sound can serve as audio-objects. A typical example is a single track of an audio CD.

4.2 Detecting Audio Objects

Detection starts with a template of an audio object. Figure 16 explains the procedure. The audio object template undergoes a set of predefined parametrized image-transformations[4] before a visual detecting algorithm is used to possibly detect the modified template in the audio track.

---

[4] These transformations, discussed in the image domain, are named image-transformation, to distinguish them from the time-frequency transformations in this chapter.

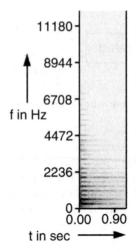

Fig. 15. Example for an audio object, which can serve as template masks: A short C2 played as an individual note by a piano. $b_{crit} = 49.13 Hz$

Sound variations, image-transformations: An audio object is represented by a visual template. Each template is transformed before comparison with the audio track. This step compensates for the different setting used for recording the template and the audio track (different microphones, different rooms, different instrument of the same kind, etc.). The different settings result in slightly different spectrograms, different levels and small sampling rate differences. While for the first and the second, the detection relies on a robust algorithm, the third can be avoided by the preprocessing step called image-transformation (see Fig. 16).

The sampling rate differences have the following impact on the template or an audio track:

- A higher sampling rate results in more samples in a given time. Compared to the correct sampling rate this results in a longer image. The template has to be compressed in time direction to compensate for this.
- A higher sampling rate recording played at the correct sampling rate results in a lower sound. The template has to be stretched in frequency direction to compensate for this.

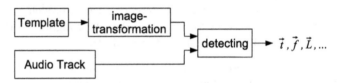

Fig. 16. Detecting audio objects

– Regardless of the sampling rate difference, which is unknown, a compression in one comes along with a stretching in the other direction. The template therefore undergoes a combined stretching and compression.

Detecting: The visual detecting algorithm is used with each possible set of allowed image-transformation parameters in order to find the best matches in the spectrograms under the allowed image-transformation space. As a result, a vector of locations (time, frequency, and shape) and perhaps other parameters such as volume level and alike are given wherever the template has been detected in the audio track.

In our system we use the normalized cross-correlation between the modified audio template image and all possible locations in the spectrogram. As a result, audio object are usually located with pixel accuracy. Since audio editing is very sensitive to inaccuracies, the estimated locations and the associated transformation parameters of the reference (i.e. template) audio object could be further refined by the Lucas-Kanade feature tracker ([216]) – an optical flow method – leading to subpixel accuracy. This technique for example is used in conjunction with sub-pixel accurate feature point localization in camera calibration (see [146]).

4.3 Deleting Audio Objects

An audio object, which has been detected in the audio track, can now be edited by the use of the best matching template. Possible editing tasks are: correcting the volume level, applying a selected equalization or deleting the sound object. In this section, the two most important methods for deleting audio objects are mentioned.

Stamping: The first approach simply "stamps" out the template out, i.e. the magnitude values are set to zero. Either a user decides interactively, by inspecting the spectrograms visually and the result aurally, or the template is applied automatically. All magnitude values, which in the template spectrogram are larger than a certain frequency dependent threshold value are stamped in the audio track (see Fig. 17).

We apply this approach to a mixed music and whistle signal. Figure 18 shows the relevant signals and spectrograms: a) music signal, b) whistling signal, c) mixed signals. Both signals can be recognized in the spectrogram and

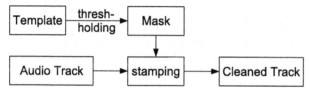

Fig. 17. Scheme for stamping a detected audio object with a template spectrogram

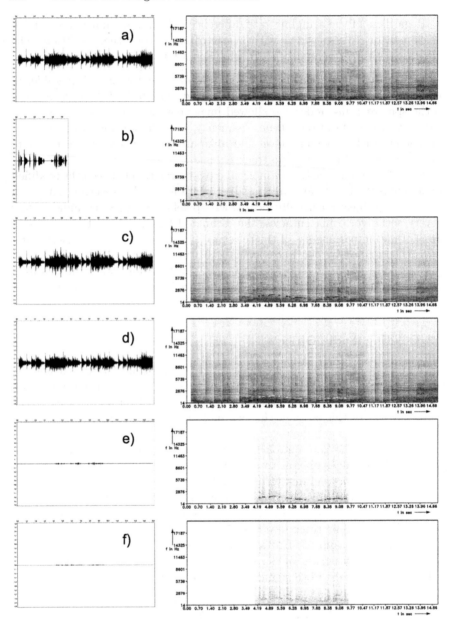

Fig. 18. From top to bottom a) music signal, b) whistling signal, c) mixed signals, whistling 4*sec* delayed, d) stamped signal, i.e. cleaned signal, e) over-compensated signal parts and f) under-compensated signal parts

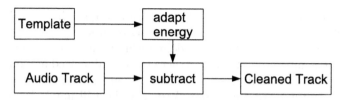

Fig. 19. Scheme for erasing a detected audio object with a template spectrogram by subtracting the magnitude values

the time delay of the whistling of 4*sec* can be determined easily. In Fig. 18, d) the "cleaned" signal is shown, i.e. the whistling is stamped out with the template spectrogram, which was created from a different recording of the same whistle signal. Figure 18 e) shows the over-compensated signal parts and Fig. 18 f) the under-compensated signal parts. In the cleaned signal the whistling is hardly perceivable and the speech signal has no perceivable difference to the uninterfered original.

Energy based erasing: In contrast to visual objects, which are often intransparent, audio objects are always additive, i.e. they shine through the energy of an other audio object. The stamping approach, although attractive because of its simplicity and analogy to the visual domain, creates poorer results for increased overlapping of objects in the time-frequency domain.

Another method is subtracting the magnitude values of the templates spectrogram from the magnitude values of the mixed signals spectrogram. As the template was recorded with a different microphone and perhaps has a different level, it is first adapted in order to match the mixed signals spectrogram absolutely and per frequency band as well as possibly before applying the difference. Figure 19 shows a scheme for erasing an audio object by subtracting the energy of an adapted template spectrogram. The success of this method depends on the similarity of template and original signal and the adaption of the template to the original signal.

5 Conclusion

We presented a new approach of editing audio in the spectrogram. It opens up the possibility to freeze audio, which is naturally transient, and edit it in a static setting. This enables many new possibilities in terms of accuracy and flexibility not only in analyzing, but also in editing audio manually and automatic. We explained how to adapt the Gabor transformation for this task in order to always reach the absolute physical limit of time-frequency resolution. This is a great advantage, because the ear itself approaches this limit very closely and is very sensitive to errors in audio signals. We presented manual as well as smart user-assisted audio brushing techniques, which are based on the use of known audio objects. Audio objects in general enhance the flexibility in achieving high quality editing results.

Future developments could be to adopt advanced image manipulation techniques, such as inpainting, to reconstruct erased or damaged audio parts and the use of artificial neural nets and other machine learning techniques to further automate the editing processes. Visual Audio will unfold its full power by combining classical techniques with presented techniques in one tool.

Part III

Algorithms II

Interactive Video via Automatic Event Detection

Baoxin Li[1] and M. Ibrahim Sezan[2]

[1] Department of Computer Science & Engineering, Arizona State University, Tempe, AZ, USA.
baoxin.li@asu.edu

[2] SHARP Laboratories of America, Camas, WA, USA.
sezan@sharplabs.com

1 Introduction

With the fast increasing amount of multimedia data in various application domains such as broadcasting- and Internet-based entertainment, surveillance and security, training, and personal audio-video recording, there is an emerging need for efficient media management including efficient browsing, filtering, indexing and retrieval. Although techniques of automated media analysis are typically employed for efficient processing, there exists the so-called "semantic gap" between the rich meaning that a user desires and the shallowness of the content descriptions that existing automatic techniques can extract from the media. In addition, in applications where a user must consume a large amount of data (which could contain more than one stream of multimedia input) in a limited amount of time, there also exists an interface bottleneck. That is to say, it is impossible for the user to apprehend all the important information in the data (let alone interact with the data) if the streams are presented in a conventional, linear fashion. These challenges call for new schemes of multimedia processing, management, and presentation.

This chapter addresses the problem of efficient media management and presentation through event modeling and automatic event detection in multimedia inputs. For efficiently managing and presenting multimedia data, we propose an event-driven scheme that organizes linear input streams into a structure that is indexed by important events defined according to the underlying application domain. This structure can then be utilized for nonlinear navigation of the data (for example), enabling not only efficient consumption of the data but also easier and more informative or entertaining user interaction. Event modeling and detection provide rudimentary semantic information through media analysis, which may be used to support content-based management of the media. However, it is in general extremely difficult, if not impossible, to rely on automatic content analysis techniques to extract all desired semantic information from raw video data. Fortunately, the proposed general modeling supports an approach to automatically synchronizing the video with independently generated rich textual metadata in order to augment the rudimentary semantic information with the information-laden textual annotation. This helps to bridge the semantic gap mentioned above.

Upon the completion of event-driven content analysis and possible synchronization of multiple streams of inputs, rich media can be presented to viewers in most convenient and effective way. To this end, a novel user interface is provided for interactive multimedia presentation. With this interface, a viewer can play back a video in various modes, such as the summary mode in which only detected key events are played back to back. Also, a viewer can choose to visualize the textual annotation at the same time the video is being played back, in perfect synchronization, achieving a truly rich multimedia experience. Further, by organizing the resulting metadata into an MPEG-7 compliant format, various modes of navigation can be supported. For example, a user can choose to play back only those key events involving a particular action. MPEG-7 standard also readily allows exchange of resulting metadata among different end-user devices and content providers, enabling personalized content or content that can be efficiently navigated. This essentially contributes to solving the interface bottleneck problem.

While the modeling and event detection methodologies that we present in this chapter should find general application in a variety of domains, in this chapter we heavily use the domain of sports video as an example in illustrating the points as there is an insatiable appetite for consuming sports video among the general population. In addition to applications to broadcast sports video, we also present a special application of the proposed scheme: an automatic video analysis tool for assisting coaching staff of football teams in effectively managing and rapidly navigating, analyzing, and interpreting training videos.

1.1 Related Work

The key task in our approach to multimedia modeling and analysis is event detection. Although not always explicitly called "event detection", considerable work on automatic video analysis has examined this problem from different perspectives. For example, in [294], Siskind and Morris proposed a method to classify six hand gestures. Siskind and his co-workers later also proposed a framework for a learner [101], addressing a similar problem, where the authors introduced a propositional event-description language called AMA. In both cases, only simple experiments (e.g., some simple actions by a single hand) were reported. The extension of such frameworks to complex events with rich visual contents is difficult, but there have been a number of studies that have attempted. In [213], a method for detecting news reporting was presented. In [289], Seitz and Dyer proposed an affine view-invariant trajectory matching method to analyze cyclic motion. In [310], Stauffer and Grimson classify activities based on the aspect ratio of the tracked objects. In [122],

Haering et al detect hunting activities in wildlife video. In [164], Izumi and Kojima described an approach to generating natural language descriptions of human behavior from a video, where the descriptions correspond to verbs characterizing the motion trajectories and three-dimensional pose of the detected human head. In addition, there are numerous papers in the literature covering topics such as visual tracking of moving objects or humans, gesture and gait related analysis or recognition, surveillance including traffic monitoring, (human) motion or activity and behavior analysis etc (e.g., see [196, 1, 340, 73, 108, 242, 186]). In specific applications such as automatic analysis of sports video, where event is relatively well-defined based on the domain knowledge of the underlying sport, there is also a large amount of work on event detection (e.g., see [182, 199, 201, 198, 272, 281, 361]).

Even with the extensive existing work, there still has been a lack of progress in the development of a unifying methodology or framework for temporal event detection, and in the development of a general metric for measuring performance. Certainly the successful development of such framework would tremendously benefit many real-world applications. The general framework proposed in this chapter attempts to address this issue.

2 Event-based Modeling and Analysis of Multimedia Streams - A General Framework

A multimedia stream (e.g., a video) recording physical activities can last an extended period of time, although there might be only a small portion of the stream that is of interest to any human operator. For example, a typical sports broadcasting program may last say, a few hours, but only parts of this time contain real actions of the underlying game. Also, a surveillance camera may continuously capture video, of which only a few shots are of interest in general. Even for personal media, the significantly informative parts will become relatively very small if the media is set out to become the lifetime store of (almost) all the experience as in the examples illustrated in [111, 82]. Although a media stream may be naturally organized by the associated timestamp, in these examples illustrated about, the timestamp alone does not lend itself to an efficient representation of the underlying structure of the media so as to support operations such as retrieval of particular media segments and nonlinear browsing.

In recognizing the fact that, when facing a long-lasting media stream such as a long sport video or a surveillance video, humans are primarily concerned with only those important moments corresponding to particular physical events, we can model the input stream as a sequence of "events" interleaved with "non-events", with "event" being defined as the basic segment of time during which an important action occurs, and thus in the case of video, we have effectively the simple two-state model as illustrated in Fig. 1.

While the model of Fig. 1 may seem to be trivial or be just stating the obvious, we will show, based on this simple model, that we will be able to

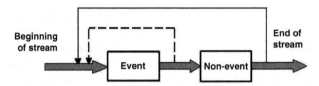

Fig. 1. A general model of a media stream in terms of "event", defined as a segment of time during which an important action occurs. The inner loop (in dashed lines) indicates the possibility that two (or more) important actions can occur consecutively

design general algorithms for the basic problem of event detection in temporal multimedia data. For example, in sports video analysis, when the model is instantiated for a specific sport, the event can be, for example, a pitch in a baseball game, an attempt of offense (i.e., a play) in football, a bout in wrestling, and a goal or a goal attempt in soccer, etc. Effectively, the general model becomes one of modeling a sport video with interleaved "play" and "non-play" segments, as shown in [198]. It was further shown in [198] that an "event" can be defined as a replay segment in modeling continuous-action sports such as soccer.

As we discuss in Sect. 5, the model defined here also facilitates synchronization and merging of independently generated rich textual metadata for a stream with events or interesting action segments since complete and detailed textual annotations of a stream typically have the fine granularity of an event or exciting action. For instance, in a soccer video, any complete textual annotation will include the description of goals and exciting goal attempts.

Obviously, this modeling is applicable to a general time-indexed stream and it is not restricted to video. Nevertheless, for convenience of discussion in the following, we will focus on primarily the case of video. However, the generality of the simple model is important to the arguments of Sect. 6, where the support of multiple cues is discussed.

3 Event Modeling via Probabilistic Graphical Models

The simple break-down of an input video into interleaved events and non-events as shown in Fig. 1 can naturally support an event-based manipulation of the video, which is key to practical needs of content-based retrieval and efficient non-linear navigation. The problem is then how to break down an input video as such, i.e., how to detect the event segments from the entire video. This issue is addressed in this section.

A temporal event may contain different components that have distinct features. We use the term *state* to represent the individual components. In a temporal event, the order at which the states occur may be critical to detecting the event. Therefore, even if the individual states have been detected, the confirmation of an event cannot be done until the temporal relationships among the states are determined. While this seems to be a trivial issue, as one might think that in temporal data the detected states are already naturally ordered by the time axis, it is not that simple in reality. For example, in many cases the detection of the states may not be completely accurate, and thus we may need to maintain multiple candidates at each time instants, resulting in multiple possibilities for time-ordered state sequences. The issue is also complicated by the fact that two distinct events may share the same set of states, and only the orders of the actions are different. (While this may not be that intuitive, it is very likely, especially due to the fact that the states are often represented with only simple cues detectable by an algorithm, and thus the ambiguity at the state level may be inevitable.) In this case, the temporal relationships between the states are in fact an additional constraint that should be exploited to filter out less likely state candidates. It is in this sense that we say that event detection often boils down to the inference of (typically) causal relations between different components (states) extracted from the sensory data.

To facilitate this inference of causal relations between individual states of an event, we propose using a directed graphical model for modeling an event, with the vertices of the graph representing states of the event, and the directed edges between vertices representing possible temporal transitions between states. Each edge is further associated with a probability, specifying how likely that transition may occur. Thus the model of Fig. 1 is expanded into the illustrative example as in Fig. 2, where we assume that there are at most five states in any one of the possible events.

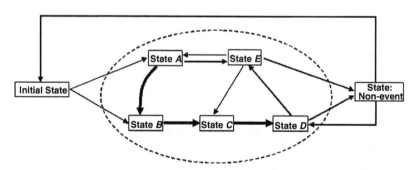

Fig. 2. The effective model of an input stream through modeling a physical event by directed graph, assuming that there are five distinct states in the sensor data. The encircled states form the event model, while the video is modeled by the entire graph. The arrows between any state pair indicate allowed transitions between states. The shown transitions are for illustration only. One particular event may not traverse all the possible states. For example the bold-lined path A-B-C-D may stand for an event occurring in the sensor data. The multiple entries into the encircled event model indicate that events may start at different states

We want to emphasize again that since in practice the states may be characterized by only low-level features with ambiguity, and thus a sequence of temporal data from the sensors may always fit nicely into some states of the model at any given instant, only the transition of states provides further constraints for signifying a particular event. Therefore, in the proposed approach, a key event will be modeled by particular transitions between a set of vertices, with each vertex typically being a state characterizing a physical situation that is a constituent component of the underlying event. Further, the transition edges are associated with a measurable feature vector $\vec{V}$ to obtain the following model in Fig. 3. These feature vectors are detected by automatic algorithms. With a probabilistic approach, the association of a feature vector with a transition edge demands a probability density to describe how likely that particular transition occurs with different values of the feature vector.

We claim that this modeling is general in the following sense: the states (vertexes) could virtually be anything that is meaningful and useful in a particular application domain and their relations are determined also by the domain. Thus it is not hard to imagine that, for example, while in video-based surveillance the states could correspond to the actions/gestures/locations of any tracked moving subject, they can very well be a few characteristic video scenes that relates to parts of a play in a sports video.

Given such a general model of Fig. 3 based on directed graph, we can design different algorithms to infer the likely sequences of states given a sequence of measured feature vectors and thus achieving the detection of events. This inference could be both deterministic (where the transitions are either allowed or disallowed) and probabilistic (where the transitions are with certain probabilities). While the modeling can certainly entertain both types of inferences (and indeed in Sect. 6 we will illustrate both types of algorithms), we can view the deterministic approach as a special case of the probabilistic approach with particular densities. Hence the modeling becomes one of a probabilistic graph model , with the edges being associated with probabilities

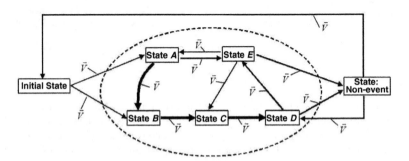

Fig. 3. The complete model, with each edge in the graph associated with a feature vector measurable from the sensor data. Not shown are the probabilities (densities) associated with the graph (including the marginal densities)

(or probability densities). A major motivation for adopting a probabilistic framework is for handling the ambiguity (noise and uncertainty). In addition, the probabilistic approach can also exploit automatic learning capabilities to infer hidden causal relations between complex input data sources, which may otherwise be extremely difficult to determine. Therefore, probabilistic approaches can lead to elegant solutions when deterministic inference cannot be easily performed in complex real world problems (this will be further illustrated in Sect. 6). The learning capabilities also allow the resultant algorithms/systems to be adaptive, i.e., the algorithms/systems can use the knowledge accumulated in previous detection to update the algorithm and system parameters.

A complete example will be discussed in Sect. 6. In the following, we illustrate how a baseball play can be modeled by a simple four-vertex probabilistic graphical model, as reported previously in [199]. Figure 4 shows sample frames from a typical baseball video, where the event is defined as a baseball play (a pitch, a run, etc.). In this case, the event detection task becomes one of play detection. A special case of the graphical model, the Hidden-Markov-Model (HMM), was used to successfully detect baseball plays, as illustrated in Fig. 5, where the feature vector is three-dimensional. The three components of the feature vector include: the similarity to a coarsely-defined model image, the average motion magnitude, and the scene transition type (details are reported in [199]).

4 Event Detection via Maintaining and Tracking Multiple Hypotheses in A Probabilistic Graphical Model

With an event modeled as above, probabilistic formulations can be utilized to describe the dynamics of the set of states, i.e., the transitions between

Fig. 4. Sample frames from a baseball video clip, which contains a few pitches. The frames are one second apart in a 30 frame per second video. The event is defined as a baseball play (a pitch, a run, etc)

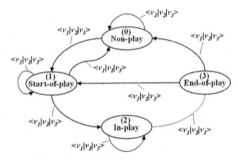

Fig. 5. A special case of the probabilistic graphical model - a Hidden Markov Model - for event modeling in a baseball video. A video is assumed to traverse between the states from frame to frame. At each frame, the feature vector is evaluated, and the value will determine the transition probability of the video from the current state to the next. The most likely transition path is then obtained. If the path conforms to an event definition (i.e., a particular sequence of states), an event is declared. See [199] for details

the states. In this setting, the detection of a key event can be achieved by examining the transition paths and the corresponding probabilities. Thus, effectively, the event detection is formulated as one of statistical inference. Formally, the above formulation and the task of event detection under this formulation can be defined as follows:

- **Event modeling:** Let a graph $G = (V, E)$, where V is the set of vertices and E the set of transition edges. Associated with each edge $e \in E$ is a random vector $\mathbf{x}_e$. For any subset S of E (think S as a particular path corresponding to a sequence of states), we define the set of random vectors $\mathbf{x}_S = \{\mathbf{x}_e | e \in S\}$.
- **Event detection:** Given a probability distribution $p(.)$ defined by the graphical model, the event detection problem can be solved by computing either the likelihood of the observed data or by computing the marginal distribution $p(\mathbf{x}_S)$ over a particular subset $S \subseteq E$.

There are established algorithms (e.g. the forward-backward algorithm and the Viterbi algorithm [259]) for solving these computational tasks, as will be illustrated later with the example of HMM, which can be viewed as a special case of the probabilistic graphical model. Formulating the event detection problem as one of statistical inference on probabilistic graphical model also allows us to adopt methods of variational inference (e.g., see [175]), in addition to the Markov chain Monte Carlo methods.

In the above graphical model for an event, following a particular (and actual) path of the event in the model is equivalent to unfolding the components of the event. In the process of observing the occurrence of an event, at the very beginning, a human or a system has a state of knowledge that is characterized by uncertainty about some aspect of the physical event. After observing some data, that uncertainty is gradually reduced. The amount of uncertainty reduced should be coded into next step of observation so that at the next step the uncertainty can be further reduced. The probabilistic graphical model entertains this particular view of event detection nicely and the progressive reduction of uncertainty in event detection is realized naturally with the model.

Using the simple model in Fig. 5 for baseball, an event detection algorithm has been reported in [199], where the model is presented as an HMM. Li et al. [199] are among the first who used HMM to model and detect sports events, where two approaches were proposed to solve the event detection problem as an inference problem. The first approach in [199] assumes that video shots have been detected and that each shot is generated with probability by certain underlying state. We consider the four-state HMM shown in Fig. 5. Training sequences of shots with pre-specified play/non-play segmentation are used to estimate the model parameters. To detect plays in an input video, one first obtains a sequence of shots. Then the most likely sequence of states is found by the Viterbi algorithm ([259]). Plays are detected by identifying sequences of states "1-2-3".

The second approach simultaneously addresses both shot-detection and high-level inference. We still use the four-state model in Fig. 5, assuming that each arc is associated with an observation vector. The algorithm works as follows. For parameter estimation (i.e., learning the model parameters), a feature vector is computed for each frame in the training sequences. Each frame in the training sequences is labeled with one of the four states. Parameter estimation for the HMM is done using Baum-Welch algorithm ([259]). With the ground truth (state labeling) for each frame given, we can compute an initial model from the training sequences, instead of using a random or ad hoc handpicked initial model so that the Baum-Welch algorithm converges to a better critical point than using a random or ad hoc initialization (see [199]). To detect plays in an input video using the trained model, the same feature vector is computed for each frame, and Viterbi algorithm is then applied to find the most likely sequence of states. A sequence of "1's-2's-3" signifies a play. Even with only such a simple feature vector as illustrated in Fig. 5, very good detection performance was reported (89% detection rate) in [199].

The above simple example is intended to show how exact algorithms can be used to solve the event detection problem within the proposed probabilistic event model. As discussed earlier, with the proposed model, much more can be done, including maintaining simultaneous multiple hypotheses, using variational inference, performing on-line learning, etc, which we will discuss in the following section.

5 Advantages of Using the Probabilistic Graphical Models

Despite its simplicity, the proposed modeling has obvious advantages both in theory and in practice, as outlined in the following.

Maintaining Multiple Hypotheses:

Note that, while an exact algorithm such as the Viberbi algorithm can provide a single solution, given the measurement and the model, it may be beneficial to maintain multiple paths in detecting an event, where each path is associated with a likelihood function. In this way, the system can output a few most possible events (assuming that the most likely paths all have corresponding events, for simplicity of discussion), and present them to a high-level analysis procedure for further processing. This may practically mimic the human behavior in second-guessing an event by dropping the first choice upon the availability of a single bit of new and critical information.

Learning Capability:

While the structure of the model may be determined based on domain knowledge, the model parameters are typically learned from labeled training data (although knowledge of human experts may be hard-coded into the model parameters as priors). The learning capability provides the opportunity for adapting the algorithm/system as it is running in real time: whenever the output of the system, which is used to guide some action, is not penalized, the system can believe that the event detection task that it has just completed is performed correctly, and thus the data of that detection can be used to update the model parameters. Thus the resultant system is one with automatic and adaptive reasoning capabilities. The adaptation/learning is typically realized by only updating the probabilities (or probability densities) in the model. This is an advantage over a deterministic inference method where adaptation may require changing some hard thresholds or even the rule base.

Multiple Input Modalities:

Up to this point, we have not explicitly addressed the issue of information fusion with multiple modalities. Nevertheless, as illustrated in Fig. 3, multiple cues are associated with each transition arc as a feature vector. There is no requirement that the cues must be from the same modality, and thus audio cues, for example, can be used as components of the feature vector. In this way, the information fusion is effectively realized through the model itself. To give a specific example, in Fig. 5, one dimension of the feature vector is the averaged motion magnitude, another is the scene cut type. These two types of cues are of different nature, although they are both computed from the visual channel. It is not difficult to imagine that in applications where a camera is used in conjunction with other types of sensors such as an inertial sensor, the extracted features from the measurements in those sensors can simply be incorporated into the feature vector (with proper normalization).

Supporting Synchronization with Other Stream:

Another important advantage of the proposed event-based modeling is the fact that it facilitates synchronization and merging with independently generated rich metadata, which typically have an event-based granularity. Upon synchronization and merging, the resultant composite metadata contain not only video indexing points of event segments and simple classification of events contained therein, but also independent rich metadata that are typically generated by human experts. This composite description stream, when applied to the underlying video in the context of database indexing and retrieval, enhanced nonlinear browsing, and summarization, significantly increases the value of the video content for both content providers and end users. In this sense, the challenging "semantic gap" is deemed as bridged. Details of work in this direction have been reported in [198].

In the case of sports video analysis, a set of algorithms, largely based on earlier developments of the general modeling, have been developed, which are collectively referred to as Hi-Impact Sports [84]. Compared with earlier work in this area, the model is more general as it can unify two major types of sports, namely action-and-stop and continuous-action sports, as discussed in [198]. For example, for action-and-stop type of sports, the proposed approach is a superset of earlier work (e.g., [182, 199, 373]), where individual papers either handle a specific sport [182], or can handle only play/break type of patterns [199, 373], which is only appropriate for action-and-stop type of sports. Also, in the case of continuous-action sports, by defining the event on the basis of replayed segments (see [198]), the resulting model has attractive advantages compared with other approaches. For instance, the work of [346] focuses on structure analysis and exciting segments are extracted based the analysis results; In [361], Yow et al. extract soccer highlights based on analysis of the image content; The work of [272] uses some audio features in extracting highlights; In [91], Ekin and Tekalp break the video into shots and then analyzes them (the analysis of the shots may or may not result in classification of the shots into important or non-important shots). In our modeling, exciting actions are defined based on the selection made by a content production expert, namely on replays. Therefore, in terms of accuracy or meaningfulness in determining whether an action is exciting, our modeling has a unique advantage over the above mentioned methods that practically rely on the automatic understanding of the content by a computer. Only the automatic detection of replay segments is based on low-level features, which is a much more suitable task for a computer than understanding the content.

6 A Case Study: Football Coaching Video Analysis

In previous sections, we have seen how the general modeling and the corresponding algorithm can be used in baseball event detection. More comprehensive results, including those for other sports including American football and soccer, can be found in [198, 200]. In this section, we focus on a simple but interesting application - American football coaching video processing and analysis [201] and show how the algorithms and results from [201] can be cast into the general technique presented in the previous sections.

6.1 American Football Coaching System

American football coaches, both of professional teams and of college teams, routinely use a specific type of video in their training and strategy planning. This video ("coaching tapes") is formed by a human operator by editing video captured from two or three different camera angles during a live game. Each camera captures all the plays of one game. For example, in the most common sideline/end-zone interleaved coaching video, each play contains a scoreboard shot (SB), followed by a sideline shot (SL) of the play and then followed by an end-zone shot (EZ) of the same play, as illustrated in Fig. 6. There are other variations of combinations. A football coach uses not only the video tapes from his own team but also tapes from other teams, and routinely logs the tapes and adds his own annotations for each play. Currently, a coach can use commercially available logging systems for viewing the coaching video and adding his annotation. The coach, however, has to painstakingly seek the start point of each play/shot before adding annotations to that play/shot. Therefore, there is a need for automatic parsing of the video in order to provide quick nonlinear indexing and navigation capabilities.

To achieve this automatically, the key task is to detect each play and its constituent shots so that the logging system can provide a coach with quick and accurate access points to each play. This appears to be a simple task at first glance, since the video is already edited. However, the challenges arise from the fact that the edited tapes contain many unusual situations, which

Fig. 6. Three constituent shots of the same play (from left to right): scoreboard, sideline shot, and end-zone shot

might be due to irregularities in the live production of the tape. For example, some plays do not have the scoreboard shot and some plays contain more than one scoreboard shot. Moreover, coaches demand 100% accuracy before they can accept any automatically-generated results. Otherwise, annotations may be associated with wrong plays. For example, a 97% accuracy, which might be very satisfactory in many pattern detection applications, is not acceptable to the coaches, according to the system vendors who sell logging systems to the coaches. This stringent demand on accuracy poses a challenge to any automatic analysis algorithms. In the following, we consider both deterministic and probabilistic approaches to overcome this challenge, based on the general framework presented in the previous sections.

6.2 Deterministic Algorithm

As discussed in Sect. 3, with a proper state modeling and choice of feature vectors, a deterministic approach boils down to the inference of a deterministic transition path in the graphic model. In the case of coaching video, the states are naturally determined by the types of the shots: SB, SL, and EZ, and the allowed transitions are shown in Fig. 7, where we have utilized the (mostly) rigid periodical patterns in a coaching video to determine the transitions.

Next, we need to identify characteristic features that categorize each of the known shot types: SB, SL, and EZ. From the example shown in Fig. 6, we observe that percentage of green pixels in a frame is a useful cue for distinguishing a SB shot from SL/EZ shots. Also, the SB shot is typically very short (2 to 4 seconds), and a SL or EZ shot can be 3 to 25 seconds long. Thus the length of the shot is another cue that we can use. Yet another cue arises from the fact that a SL/EZ shot usually contains action and thus the last frame of the shot is typically drastically different from the first frame of the same shot, while a SB shot is relatively static. Therefore, the difference between the first frame and the last frame of the same shot can be used as one cue in the shot classification. In a sense, this difference reflects a shot's motion complexity, which is a commonly-used cue. A rule that governs the underlying semantics is that a play is typically an ordered triplet of (SB-SL-EZ). This fixed order provides a semantic constraint. For instance, a SL should

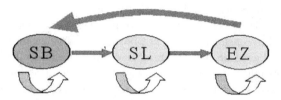

Fig. 7. A simple graph model for the three states with allowed transitions

be preceded by a SB and followed by an EZ. We make the following definitions and then design a straightforward rule-based algorithm (Algorithm I).

Algorithm I: *Deterministic Inference*

Loop for the video {

1. Detect a shot S_i.
2. Compute P_g, D, and L. If P_g, D, and L exceed pre-determined thresholds, classify S_i as a SL or EZ; otherwise classify S_i as a SB shot.
3. If S_i is a SL/EZ, check the ID of the previous shot S_{i-1}. If S_{i-1} is SB or EZ, then classify S_i as SL; if S_{i-1} is SL, then classify S_i as EZ.

}

With, P_g represents the percentage of green pixels in the first few frames of the shot; D represents the color histogram difference between the first frame and the last frame of the shot; and L is the length of the shot.

Notice that one could also try to use additional cues such as an estimated camera angle (e.g. through field line detection) for further distinguishing an SL shot from an EZ shot, and thus improve Step 3. Also, in a practical implementation, we can have many other varieties of Step 2. For instance, we can set multiple thresholds for each of those three features and perform refined reasoning to account for some irregular situations. For example, even if a shot is shorter than the length threshold for SL/EZ, but if the frames are almost totally dominated by green pixels, then it should still be classified as SL/EZ. What all these tricks do is to incorporate more sophisticated rules into the deterministic inference step. If the number of rules grows significantly, the inference becomes more and more complex. (In general, if the number of states becomes large, the problem may be intractable.)

As demonstrated in prior work (e.g., [19, 182, 199, 373]), deterministic approaches can be very successful in practice. Indeed, for this relatively simple task of coaching video indexing, the deterministic algorithm presented above works reasonably well even if it does not achieve 100% accuracy (a comparison will be provided in Sect. 6.4). They are also easy to implement and computationally efficient. However, there are two major disadvantages. First, implicitly setting the inference rules may not be easy in some cases, especially when the inference is based on a large number of cues. Second, one has to choose some hard thresholds (e.g., a threshold for the dominant color ratio in [373]). A fixed threshold cannot possibly cover the variations in real-world video (e.g., some field can have very yellowish grass or contains a special-colored logo overlay which renders the field far from being green).

6.3 Probabilistic Algorithm

The above simple algorithm will work well when the video is always as regular as shown in Fig. 6. But in reality there are many irregularities. For example, Fig. 8 illustrates plays from some other videos where the dominant green color assumption is invalidated by various reasons. In addition, the length is not always a reliable cue in distinguishing SB from SB/EZ, as shown in Fig. 9. Further, since many plays are short and may lack sufficient action (and the camera may shoot from a long range), the resultant D could be as small as, or even smaller than, that of SB shots, as illustrated in Fig. 10. These anomalies render it very difficult to achieve 100% accuracy using a rule-based reasoning as in Steps 2 and 3 in Algorithm I. In fact, with multiple cues, setting rules and choosing thresholds becomes increasingly intricate as the number of irregular cases grows with each new sequence added to the test.

To alleviate these issues, we have found that the probabilistic inference based approaches are more appealing. Specifically, while still using the simple model of Fig. 7, we now view it as a special probabilistic graph model: a simple first-order Markov transition model. As discussed in Sect. 3, in such a modeling, the transitions are not deterministic, and the problem becomes one of determining the most probably path of transitions, given the measured

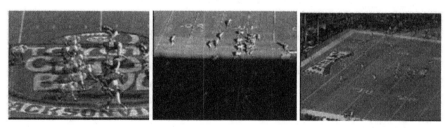

Fig. 8. Irregular cases (logo overlay on the field, shadow, distorted color, etc) that render it difficult to use the percentage of green pixels in detecting an SL/EZ shot

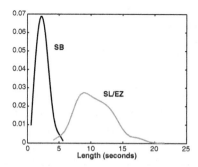

Fig. 9. Empirical length distributions of SB and SL/EZ, respectively. The overlap prevents a 100% classification using only the length information. Note that classification of SB and SL/EZ is only part of the job

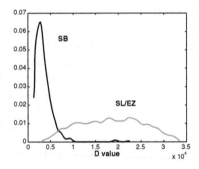

Fig. 10. Empirical D distributions for SB and SL/EZ, respectively, show significant overlaps

feature vectors. In this case, a simple algorithm can be designed to find the most probable path, as detailed below.

Let $P(Z|C)$ be the probability of observing the feature vector Z given shot class C, where Z consists of three components: P_g, D, and L. C takes either of the three labels SB, SL, and EZ. For simplicity, we assume that the components of Z are independent of each other, and thus we have

$$P(Z|C) = P(P_g|C)P(D|C)P(L|C).$$

Now, the individual probabilities $P(P_g|C)$, $P(D|C)$, and $P(L|C)$ can be set empirically or learnt from the data (e.g, using the empirical distributions in Figs. 9 and 10). An alternative is to set or learn the joint distribution $P(Z|C)$ without the independence assumption, which would require a larger amount of training data since one needs to estimate the three-dimensional distribution jointly. With a first-order transition matrix $P_t(S_i|S_{i-1})$ specifying the transition probabilities of Fig. 7, we propose the following algorithm (Algorithm II).

Algorithm II: *Probabilistic Inference*

Determine the class C_{k_0} for shot S_0.

Loop for the video {

1. Detect a shot S_i.
2. Compute Z, and $P(Z|C_k)$ for all k.
3. Compute $P(Z|C_k)P_t(S_i|S_{i-1})$, and set current shot to C_{curr}:
   $C_{curr} = argmax_{(C_k)} P(Z|C_k)P_t(S_i|S_{i-1})$

}

Apparently, this algorithm can be viewed as a simplified version of the general inference algorithms described in Sect. 2. In particular, Algorithm II can have the following Bayesian interpretation. The problem is to find the most probable state label $C_k(S_i)$ for the current shot S_i, given a feature vector Z

and the state label for the previous shot S_{i-1}, i.e, to maximize $P(C_k(S_i)|Z)$ given $C_j(S_{i-1})$. Applying Bayesian rule yields that $P(C_k(S_i)|Z)|C_j(S_{i-1})$ is proportional to

$$P(Z|C_k(S_i))P_t(S_i|S_{i-1})|C_j(S_{i-1}).$$

Thus, if no prior information is available, the solution is found by maximizing $P(Z|C_k(S_i))$. Otherwise, when $C_j(S_{i-1})$ is given, we maximize

$$P(Z|C_k(S_i))P_t(S_i|S_{i-1})$$

with respect to k to find the solution. Note that this algorithm is a simplified case of a full-fledged HMM algorithm. It can be easily extended to a full HMM algorithm similar to that in Sect. 3.

6.4 Experimental results and performance comparison

With the above algorithms as the core, we developed a product prototype called "HiMPACT Coach" [201]. HiMPACT Coach processes coaching videos on line and in real time. The system automatically identifies the in and out points of all scoreboard, sideline view, and end-zone view shots of plays in a football coaching video where, for example, sideline and end-zone views of a play are laced and corresponding scoreboard shots are placed in between plays. The system has been tested and evaluated on-site by a sports A/V systems integration company, in a configuration illustrated in Fig. 11. In the following, we briefly present some illustrative experiments using four videos from different teams (and games).

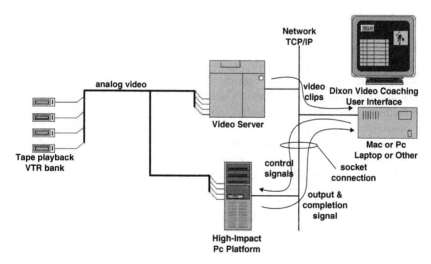

Fig. 11. A real-time testing configuration, where "High-Impact PC Platform" runs the proposed algorithm from live feed of tape players and then pass the results to other video editor interfaces (professional software platforms) for visualization

The system captures the data from a tape player using a consumer grade capture card. The data can be processed both live and off-line. In the off-line mode, the data is first MPEG-compressed. The resolution of the input to the system is set at only 120 × 160. All these suggest that the system is robust enough so that it does not demand very high-quality video input. In the off-line case, the system can process about 180 frames per second on a Pentium-4 2.0G CPU PC, with either of the proposed algorithms. Table 1 shows sample results for four testing sequences. As shown in the table, the deterministic algorithm I performs well (consistently above 94%), but it has difficulty achieving the 100% performance for all but one sequence. The probabilistic algorithm II was able to obtain perfect detection for all the test sequences.

7 A Novel Interface for Supporting Interactive Video

As mentioned previously, the event-centric modeling of the multimedia stream is intended for efficient management and consumption of a large amount of multimedia contents. With the events in the input streams detected, and especially with the multiple streams synchronized, we now turn to the task of efficient presentation of the streams to a user.

Intuitively, even if the raw inputs are of a huge amount, the detected events should be sparse in typical applications. For example, in a surveillance video, one-day-worth of video may result in only a few minutes of note-worthy video clips. Thus by simply presenting the detected events, we have achieved the goal of efficient consumption of the data. In a practical problem, only keeping the detected events may not be sufficiently useful, as a user may prefer to do more exploration of the original data after intrigued by some of the detected events. In addition, in examples where multiple streams of different modalities are present, to visualize all the data streams demands careful consideration.

In this section, using sports video synchronized with independent rich metadata as an example, we describe a novel interface for supporting interactive multimedia viewing. Specifically, we have incorporated the visual events and the synchronized metadata into an MPEG-7 compliant prototype browser featuring novel user interface paradigms for viewing sports summaries (we demonstrated an earlier version of this system in [197]). The rich media description is expressed in an MPEG-7 compliant XML document [329]. The prototype system provides a synchronized multimedia viewing experience through the implementation of a video-driven scroll window, which shows the

Table 1. Correct classification rate for four testing video.

Tape name (teams)	Algorithm I	Algorithm II
Ram-49ers	100.0%	100.0%
Cardinals-Raiders	94.2%	100.0%
Ram-Eagles	95.1%	100.0%
Cardinals-Chiefs	97.0%	100.0%

textual information corresponding to the video segment that the viewer is watching. A screen shot of the system is shown in Fig. 12 for football. The text window below the video playback window contains textual annotations that are rolled upward in synchronization with the video playback. The annotation for the event that is currently played back in the video window is displayed in gold color font. The annotation above the golden text is for the previous play that has been played back. The annotation below is for the next event. The game information at the right hand side of the rolling text window is also synchronized with the video playback and updated accordingly.

Note the semantic richness of the annotations in Fig. 12. For example, the type of the action and the name of the player(s) involved in the action are included. Clearly, the media description that is used in the enhanced browsing example shown in Fig. 12 can be used in indexing the video in a database. Due to the semantic richness of the description, users will be able to capture rich meaning as they query the database. Due to the segment structure of the description, users are also able to retrieve event clips from different games stored in the database according to their queries, e.g., using a favorite player's name. Such retrieval and browsing applications can be used in a production environment, or they can be offered as a service, e.g., on the Internet.

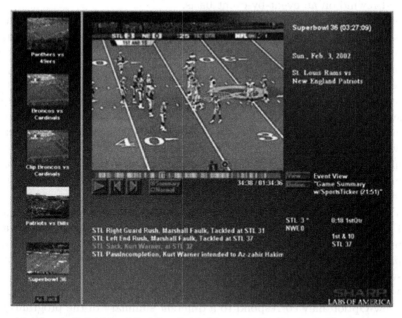

Fig. 12. Synchronized multimedia-viewing experience for football - a screen shot from our prototype system. To the right hand side and lower-right corner of the main video window, shown are the game statistics extracted from the SportsTicker data feed (the statistics are updated as the video plays). Directly below the window is a list of scrolling textural descriptions of the underlying game, with each line corresponding to a play, and the line in golden font corresponding to the current play that is being played

Key components of the user interface shown in Fig. 12 are described below.

-   Navigation Bar: The timeline of the program is displayed in a horizontal scroll bar immediately below the video window. The full width of the bar is scaled to represent the Total Time of the program. A sliding cursor box depicts the current time as a relative position along the scroll bar. The user may drag and move the cursor box to begin playback at any point. The scroll bar is shaded to show which segments of the program are part of the currently playing video summary (View-Option). Light-color regions correspond to program segments that are included in the playback (i.e., key event segments), and dark-colored regions correspond to remaining program segments that are bypassed during summary playback.

-   Next/Previous Event Advance: The user may click these buttons (to the right of the Play button below the Navigation Bar) to skip forward or backward in the program from the current cursor position. When in Summary playback mode (see "Playback Mode" below), a skip will move the program position to the next or previous program segment of the summary. When in Normal playback mode, a skip will move the program a fixed amount of time (e.g. 15 seconds) forward or backward in the program. In Summary mode, these buttons provide non-linear fast forward and rewind on an event-by-event basis.

-   Summary/Normal Mode: A user may click this button (to the right of Next/Previous Event Advance buttons) to disable/enable the summary playback mode. In summary mode, the program will be played back by presenting only the program segments corresponding to key events (represented as the light regions in the scroll bar). In normal mode, the program will play the entire program straight through to the end. When the user switches from normal mode to summary mode, the program position will advance to the next program segment if the current position is not already within a summary segment. While in summary mode, the user is able to move the scroll bar cursor to any position, even outside of program summary segments. However, when the program playback enters a summary segment, the summary sequence will be resumed.

-   View Mode: A variety of views may be available for the selected program. Different views correspond to different ways of navigating programs and program summaries. The currently active view for the program is displayed here. For example, Event View indicates that the current program summaries contain interesting or important events in the program.

-   View Option: The active view may have a variety of options available. Each View-Option may correspond to a different summary of the program. For example in Fig. 12 the current View-Option is called "Game Summary w/SportsTicker (21:51)", which indicates a summary of the sports game with a duration of 21 minutes and 51 seconds where summary segments are synchronized with SportsTicker data (generated by SportsTicker Inc.). The summary will denote how the program is to be played back, including the playback sequence and associated annotation information. The label of the currently active View-Option is displayed here.

– Preference-Based Filtering: The system allows the user to identify himself/herself to the system. When/if the user's profile is stored in the system, the system will filter and rank-order the video content stored in the system on the basis of user's profile (e.g., on the basis of favorite types of sports) and choose and order the video thumbnails at the left vertical pane accordingly. In addition, the system may provide personalized summaries, e.g., the plays where user's favorite pitcher is pitching. The extent of personalization depends on the extent of information contained in the user profile as well as the extent of the metadata associated with the video.

– Manual Annotation: Current system implementation displays independently generated and automatically generated SportsTicker data in a manner synchronized with the video events. The system can be extended to accept textual annotation from the user. The user can click on the light-green event bars to open up a text window and annotate that particular event. Manual annotations are then played back along with the video events.

Obviously, such a visualization scheme will find application in other domains. For example, a surveillance system may benefit from such a scheme. Such a surveillance application, the "favorite list" on the left side bar can be replaced by automatically selected video streams that have significant activities/events, while all the other buttons and functionalities described above can be used in efficient navigation of the surveillance video. For instance, in on-line mode, the use will be able to go back to a previous event for a quick review; in off-line "after-the-fact" searching of a large surveillance video database, the detected events (light green bars) can bring the operator immediately to the right context of the video.

8 Summary and Future Work

Although the framework and methodologies described in this chapter are general, current study is largely based on the special case of sports video. It will be interesting to see its actual application in problems such as video-based surveillance, in which tasks such as quick review of a large amount of archival video, automated annotation generation, efficient presentation that supports interactivity (e.g., even-based forward/backward searching etc) are expected to be handled well by the general framework and the visualization scheme. This is our ongoing research.

While the event-driven approach for interactive video is attractive and existing work appears to demonstrate that this is a promising method, automatic event detection in general remains to be a daunting challenge. Among the difficulties, one fundamental issue is the lack of a universal definition for event in different application domains. For example, when extending the work presented in this chapter beyond sports to other domains, one immediate question we need to answer is how to algorithmically define an event - the basic unit of the modeling. It is hard to imagine that, for example, a fixed set of low-level visual features will be sufficient to define events in various application domains. Instead of attempting to find a universal event definition, one of the authors is currently working on human-cognition-driven event modeling, in recognition of the fact that humans are in general able to quickly adapt the abstract concept of "event" to a variety of domains. In particular, we are trying to answer the following basic questions in human event detection :

– Temporal decomposition of an event: What is the most natural way of decomposing a temporal event of nontrivial complexity into its constituent components (or temporal states)? For a complex event, this decomposition will most likely facilitate the automated detection since the problem now becomes detection of simpler event components, which presumably have more coherent sensory features, and thus are easier to detect by an automated method.
– High-level knowledge versus low-level stimulus in event perception: For a given event in the sensory data, how much high-level knowledge is needed for human to detect the event, and how much of the detection is achievable using only low-level processing? For a given application, if we can separate high-level processing from low-level processing, we will be able to factor that into the design of an automated system.
– Temporal sampling of an event: For many events, a human does not require a dense temporal sampling of the event (e.g., 30 frame per second in a video) to detect them. Often, by taking casual glimpses at some "critical" parts of an event, a human can still make the detection. But how sparse this temporal sampling could be or how the temporal information might be more efficiently compressed for an event in a given application has not been fully investigated. We deem this aspect as important since it may help us to understand what the essence of an event is. Furthermore, understanding this enables us to design low-computation sensor algorithms, since the sensor needs to work (for both acquisition and processing) only at the lowest rate that still ensures the detection of the event. This problem will be naturally linked to the first problem, the temporal decomposition of an event.

We believe that we can benefit greatly from a well-thought-out study of human behavior in the process of detecting an event, especially a study that focuses on the above-listed aspects of event detection in humans. Such a study will in turn contribute to solving some of the fundamental challenges in automatic event detection.

Acknowledgement

We would like to thank *SportsTicker* and *ESPN* for providing us the textual and video data. Thanks are also extended to *Michael Dixon* of *Dixon Sports Inc.* for providing the coaching video for our experiments.

Bridging the Semantic-Gap in E-Learning Media Management

Chitra Dorai

IBM T.J Watson Research Center, New York, USA.
dorai@us.ibm.com

1 Introduction

Audio and video have become standard data in both consumer and business worlds alongside conventional text, numeric, and coded data in the recent years. As a standard data type, digital media information has become essential not only for personal communications, but also for business-to-employee, business-to-business, and business-to-consumer applications.

Business media now consist of company information, customer data, and media products as combinations of audio, video, images, animations, and other unstructured data. Collectively, these data types are known as rich media. By themselves or together with other data, they add rich detail to the description of products or services; they aid as recorded evidence of the real world in data analysis and reasoning; and they help computational tools become better adapted for human use. A few examples of business media applications include Webcasting for corporate communications, dynamic display boards for marketing with customized media and images, and voice, data, and video over IP.

Business owners and managers are starting to view business media as assets. These assets are becoming mission-critical data for many content-centric and content-rich business applications. Because the amount of media content on company intranets is exploding, business media assets retained for reference and value are growing at a much faster pace than traditional data. Business media create value by expanding the functional and business scope of applications. Business media allow companies to rapidly repurpose valuable assets (e.g., brand asset management), as well as improve collaboration, communication, and efficiency that give them a competitive edge. To reap these benefits, digital media technology must become an integral part of the company's IT processes.

Digital media content is complex and, while it might be inexpensive to create with today's technological advances, it is still generally expensive to manage and distribute. Extracting business value from media can be difficult, depending on the application. Often, business media content is used in combination with traditional business data such as training records or inventory and invoice data. to enhance customer care and realize new value.

Let us look at a few scenarios that show how enterprises are using digital media to enhance their employee and customer communications, improve the effectiveness and efficiency of business processes, and create new business and revenue opportunities.

1.1 Streaming Media Business Applications

Timely dissemination of corporate information to employees, business partners, and customers is emerging as a critical success factor for global enterprises. Digital media has proven to have high impact and improve the retention of the messages transmitted. Often, the communication has to extend across multiple cities, countries, or even continents, and must support reliable, affordable, and private desktop-to-desktop communications for effective collaboration.

Live streaming broadcasts via corporate intranets let executives deliver critical business, financial, and personnel information across the entire organization. For speedy and effective product launches in the market, streaming media-based training applications reduce the time taken to educate the marketing and sales force, enable the channels, launch the products, and inform potential customers. Voice- and video-conferencing delivered to desktops enrich the level of communication and interactions provided by email and instant messaging and act as a viable alternative to face-to-face meetings that often incur high travel expenses. The regular use of digital media enhances collaboration and interactive efforts while reducing costs. We see an evolution from earlier ad-hoc intranet streaming media deployments in companies to enterprise-wide content distribution networks for controlled streaming media delivery that better manages increased network traffic. Centrally managed business media repositories can create knowledge assets out of transient communications such as Webcasts and support their effective reuse.

1.2 Customer Relationship Management Applications

Enterprises increasingly require the capability to integrate structured data resulting from transactional applications such as enterprise resource planning and supply chain management with related, unstructured digital media such as graphics, sound files, video files, blueprints, and schematics as well as emails and document correspondences. Companies also need to improve customer interactions as part of their customer relations management.

For example, an important aspect of effectively managing a customer relationship is the ability to relate a voice conversation with a business transaction. In a typical call center today, customer calls are routinely recorded on tape, manually indexed, and archived in large tape libraries. Likewise, in investment banking and other financial corporations, interactions between analysts and businesses such as conference calls are tracked and recorded as audio or video for regulatory compliance. If information discussed in a previous call is needed, it is slow and cumbersome today to find the recording of the call.

Digitizing this information and recording it digitally along with associated metadata and business data will allow these data to be easily searched, retrieved, and shared. Integrating and correlating this business data with media data can facilitate customer service and satisfaction. This enhanced user experience can lead to increased sales and brand loyalty. Indexed recording of calls and conferences will also support enhanced information mining. For instance, financial services companies can analyze trades executed based on information callers have received.

1.3 E-Learning Media Applications

Rapid adoption of broadband communications and advances in multimedia content streaming and delivery have led to the emergence of Web-based learning as a viable alternative to traditional classroom-based education. The Internet has changed the way universities and corporations offer education and training. It propels them away from conventional means for remotely distributing and delivering courses and classroom lectures such as television networks, and steers them toward Web-based offerings.

Online learning, often called e-learning, has become an accepted means of employee training and education in enterprises because of its compelling advantages in personalization, focus, ability to track, and reduction of travel cost. Streaming media over intranets and the Internet can deliver e-learning on a variety of topics any time to accommodate the learner's schedule and style. Solutions are therefore, needed to make business-learning content management faster and easier to deploy across an organization, and to make learning content purchase and delivery efficient across businesses. Recently, researchers in both multimedia and education have focused on building systems such as e-Seminar [312] and the Berkeley Multimedia Research Center lecture browser [270] for classroom video acquisition, distribution, and delivery. Elsewhere, Rui et al. [273] discuss the technology and videography issues in lecture capture and distribution.

Although we can automate media acquisition and distribution with such systems, challenges remain in the areas of educational and training video access, standards-compliant content annotation for search and sharing, and easy sequencing and interaction with the learning media content. Automated content indexing and annotation of learning media become key tasks in customized content delivery and consumption. Particularly in the context of e-learning, there is a large amount of material such as slides, whiteboard contents, and simulations associated with the audio and video streams of lectures.

A useful learning media management system must enable cross-referenced access to all the materials pertaining to a course's media in a synchronized manner. It should also facilitate searching for a topic, a slide, or a specific instance of captured whiteboard content within a course. Finally, it should support smart browsing interfaces based on the content of the captured media data and assist in semantically guided navigation via automatically structured and tagged course media.

Automatic indexing of learning media is a key technology needed for creating useful, easy-to-use access structures beyond today's simple storyboards. We still need research on parsing and structuring media content that automatically establishes semantic relationships between the content's various elements (such as audio, slides, whiteboard, and video). We also need to develop joint audiovisual algorithms to deliver concise multilayered, concept-oriented content descriptions for search, in contrast to the low-level features such as shots, keyframes, or keywords from speech that are often used today. Finally, to foster sharing and reuse of learning content we must support data models that comply with e-learning standards such as the IEEE Learning Objects Metadata [154], Sharable Content Object Reference Model (SCORM) [6], and MPEG-7 [220].

2 Media Computing for Semantics

With the explosion of online entertainment media and business media, content management remains a fundamental challenge. Content management refers to everything from ingesting, archival, indexing, annotation, and content analysis to enable easy access, search and retrieval, and nonlinear browsing of many different types of unstructured data such as text, still images, video and audio. One of the open fundamental problems in multimedia content management is the semantic gap — that beleaguers and weakens all automatic content annotation systems of today — between the shallowness of features (content descriptions) that can be currently computed and the richness of meaning and interpretation that users desire to associate with their queries for searching and browsing media. Smeulders et al. [301] call attention to this fact when they lament that while "the user seeks semantic similarity, the database can only provide similarity on data processing". This semantic gap is a crucial obstacle that content management systems have to overcome in order to provide reliable media search, retrieval, and browsing services that can gain widespread user acceptance and adoption. There is a lack of framework to establish semantic relationships between the various elements in the content for adaptive media browsing since current features are merely frame-representational and far too simple in their expressive power. Thus, there is a serious need to develop algorithms that deliver high-level concept-oriented, functional content annotations that are parsimonious as opposed to verbose low-level features as computed today.

Solving this problem will enable innovative media management, annotation, delivery and navigation services for enrichment of online shopping, help desk services, and anytime-anywhere training over wireless devices. Creating technologies to annotate content with deep semantics results in an ability to establish semantical relationships between the form and the function in the media, thus for the first time enabling user access to stored media not only in predicted manner but also in unforeseeable ways of navigating and accessing media elements. Think about the innovative use, relational databases led to when they were introduced! Similarly, semantics-based annotations will break the traditional linear manner of accessing and browsing media, and support vignette-oriented viewing of audio and video as intended by the content creators. This can lead to new offerings of customized media management utilities for various market segments such as education and training video archives, advertisement houses, news networks, broadcasting studios, etc.

In an endeavor to address these underlying problems, we have been both advocating and practicing an approach that markedly departs from existing methods for video content descriptions based on analyzing audio and visual features (for a survey of representative work, see [301]). We have demonstrated that to go beyond representing just what is directly shown in a video or a movie, the visual and emotional impact of media descriptions has to be understood. Our contention is that both media compositional and aesthetic principles need to be understood to guide media analysis.

What avenues do we have for analyzing and interpreting media? Structuralism, in film studies for example, proposes film segmentation followed by an analysis of the parts or sections. A structuralistic approach to media computing is evident in recent emphasis on computing the underlying "meaning of the established relations between the single components of a multimedia system and exposing the main semantic and semiotic information hidden in the system's unified structure" [232]. Structural elements, or portions of a video, when divested of cultural and social connotations can be treated as plain data and therefore, can be studied using statistical and computational tools.

Another rich source is production knowledge or film grammar. Directors worldwide use accepted rules and techniques to solve problems presented by the task of transforming a story from a written script to a captivating visual and aural narration [14]. These rules encompass a wide spectrum of cinematic aspects ranging from shot arrangements, editing patterns and the triangular camera placement principle to norms for camera motion and action scenes. Codes and conventions used in narrating a story with a certain organization of a series of images have become so standardized and pervasive over time that they appear natural to modern day film production and viewing. However, video production mores are found more in history of use, than in an abstract predefined set of regulations, are descriptive rather than prescriptive, and elucidate on ways in which basic visual and aural elements can be synthesized into larger structures and on the relationships that exist between the many cinematic techniques employed worldwide and their intended meaning and emotional impact on movie audience.

2.1 Computational Media Aesthetics

Media aesthetics is both a process of examination of media elements such as lighting, picture composition, and sound by themselves, and a study of their role in manipulating our perceptual reactions, in communicating messages artistically, and in synthesizing effective media productions [365]. Inspired by it, we have defined computational media aesthetics [87] as the algorithmic study of a variety of image and aural elements in media founded on their patterns of use in film grammar, and the computational analysis of the principles that have emerged underlying their manipulation, individually or jointly, in the creative art of clarifying, intensifying, and interpreting some event for the audience.

Computational media aesthetics enables distilling techniques and criteria to create efficient, effective, and predictable messages in media communications, and to provide a handle on interpreting and evaluating relative communication effectiveness of media elements in productions through knowledge of film codes that mediate perception, appreciation and rejection. Whilst the area of affective computing aims to understand and enable computers to interpret and respond to emotional cues of users, the new field aims to understand how directors utilize visual and sound elements to heighten the emotional experience for the audience. Computational media aesthetics exposes the semantic and semiotic information embedded in the media production by focusing not merely on the representation of perceived content in digital video, but on the semantical connections between the elements and the emotional, visual appeal of the content seen and remembered. It proposes a study of mappings between specific cinematic elements and narrative forms, and their intended visual and emotional import.

Why does this research direction differ from existing approaches in multimedia processing? While others have sought to model very specific events occurring in a specific video domain in detail, this work attempts to understand the "expressiveness" of the visual and aural patterns in the medium and the thematic units (high-paced section, tranquil scene, horror shot, etc.) highlighted by them that are pervasive regardless of the specific details of the story portrayed. The underlying goal is to develop analytical techniques founded upon production knowledge for film/video understanding, to enable the extraction of high-level semantics associated with the expressive elements and narrative forms synthesized from the cinematic elements, and to illustrate how such high-level mappings can be detected and reconstructed through the use of software models. Film grammar has been used as a compositional framework in research related to content generation, synthesis of video presentations, and virtual worlds [76]. Our research, on the other hand, highlights the systematic use of film grammar, as motivation and also as foundation in the automated process of analyzing, characterizing, and structuring of produced videos for media search, segment location, and navigational functions.

We have created a framework to computationally determine elements of form and narrative structure in movies from the basic units of visual grammar namely, the shot, the motion, the recording distances, and from the practices of combination that are commonly followed during the audiovisual narration of a story. We first extract primitive computable aspects of cinematographic techniques. We then propose that new expressive elements (higher order semantic entities) can be defined and constructed from the primitive features. However, both the definition and extraction of these semantic entities are based on film grammar, and we formulate these entities only if directors purposefully design them and manipulate them. The primitive features and the higher order semantic notions form the vocabulary of film content description language (see Fig. 1).

2.2 Understanding Semantics in Media

Why is computational media aesthetics important? It embarks on a computational understanding of the use of visual and aural elements in TV and film for building tools that facilitate efficient, effective, and predictable transformation of ideas into messages perceived by viewers. Understanding the dynamic nature of the narrative structure and techniques via analysis of the sequencing

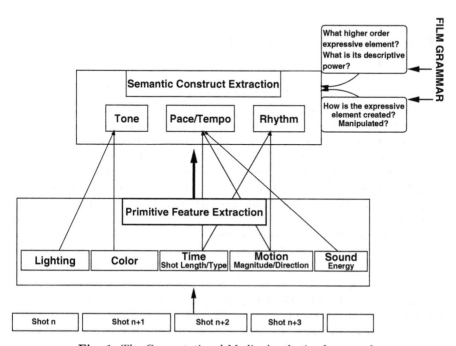

Fig. 1. The Computational Media Aesthetics framework.

and integration of audio/visual elements leads to better characterization of the content and the form than is currently possible and enables developing tools for democratic adoption of the successful techniques of the craft of film/video making. This understanding can lead to high-level semantic characterization of movies and videos and to a more effective user-query interpretation facilitating easier, human-oriented content specification in queries with media search engines. This research explores and associates deep semantics from they way they are constructed via audiovisual imagery in movies and television.

In seeking to create tools for the automatic understanding of film we have stated the problem as one of faithfully reflecting the forces at play in film construction, that is, to interpret the data with its maker's eye. We defined film grammar, the body of film literature that effectively defines this eye, and proceeded to take an example of carrying one aspect of this grammar from literature to computable entity, namely tempo and pace for higher level analysis of movies.

Film is not the only domain with a grammar to be exploited. News, sitcoms, educational video etc. all have more or less complex grammars that may be used to capture their crafted structure. In this brave new world of self-expression, soon there will be a need to manipulate digital aesthetic elements to deliver messages in many different ways. Computational media aesthetics takes us towards this goal. Computational media aesthetics allows one to learn from the practitioners of artistic expression, to design tools and technologies for the non-linear manipulation of media elements.

3 Computational Media Aesthetics at Work: An Example

The new media production knowledge-guided semantic analysis has excited many researchers who are frustrated with continued focus on low level features that cannot answer high level queries for all types of users, and they are applying the principled approach of computational media aesthetics to analyzing and interpreting diverse video domains such as movies [3, 4], instructional media, etc. In the remainder of this chapter, we describe as an example of this process our software model of various narrative elements in instructional media including discussion sections, slides, web pages, and instructor segments.

Instructional video is sometimes defined as a "motion picture designed to teach" [142]. This separates instructional video from other kinds of motion pictures owing to its explicitly stated purpose. An instructional film can fall into several categories such as documentary films, classroom lecture videos, or training videos. In [142], Herman presents a comprehensive classification of this film genre. In this chapter, we are interested in analyzing *industrial* and *classroom* educational films. Industrial instructional video refers to material used in industrial settings for various training purposes such as how to operate a machine or how to address safety issues in the workplace. Classroom teaching films are those that capture classroom lectures of lessons on various subjects.

In the following sections, we describe commonly observed patterns and conventions in instructional media productions to structure and manipulate the presentation of the learning content. We exploit production grammar to shape our understanding of the common structural elements employed in instructional videos and outline techniques to extract the narrative structures observed.

3.1 Narrative Structures in Instructional Media

The structure of an instructional video, just like any other video is influenced by the nature of information being presented and the purpose for which the video is used [170]. Instructional videos aim to 'educate'. A well-produced video that motivates its audience to gain knowledge or enables them to learn or refresh a skill is not easy to create. Decisions are made by content producers about not only what is to be shown but also how it is to be presented. Based on the narrative flow widely used in educational video for pedagogical reason, we observe a hierarchy of narrative structures which are considered as critical elements in presentation.

At the top level of the hierarchy, there can be four main categories: *on-screen narration, voice over, linkage* sections and *discussion* sections. At the next level, these four categories have further segmentation.

On-screen narration sections refer to segments in the video where narrators are shown. The purpose of these sections is for the narrators or instructors to address the viewers with a voice of authority and such sections are used to introduce a new topic, to define a concept or to guide the viewers through some procedure with examples. We further distinguish on-screen narration into two finer categories.

- *Direct narration.* This involves segments where the narrator faces the audience and where the narrator speaks to the viewers directly. The face of the narrator usually is shown in the beginning of the segment and there is little movement through the section. We refer to them also as instructor segments.
- *Assistive narration.* Here, although there is the presence of the narrator, the attention of viewers is *not* necessarily focused on the narrator. The purpose is to *emphasize* a message additionally by the use of text captions and/or to let the audience *experience* the lesson via relevant scenes with music in the background or with other environmental audio. Typical examples for this structure are shots of the narrator walking around in the factory floor and explaining the safety issues.

Voice over sections are those where the audio track is dominated by the voice of the narrator *but without* his or her appearance in the frames. The purpose of these segments is to communicate to the viewers using the narrator's voice and supplement it by pictorial illustrations or other demonstrations as visual information. Voice over sections are further grouped into:

- *Voice over with informative text.* Along with the voice of narration, there is only superimposed text on the screen. The text is usually presented on a simple background. An example of this segment are slides shown in a classroom lecture video.
- *Voice over with scenes.* This is the most often encountered structure in instructional videos, and it contains voice over in the audio channel and general scenes as visuals, but there is no text caption displayed.
- *Voice over with text and scenes.* This structure captures voice over sections where both text and scenes are found in the visual channel.

Linkage sections are in the video to maintain the continuity of a lesson or a subject. We subdivide this category into the following two groups:

- *Text linkage sections* contain transition shots encountered during the switching time between one subject to the next. Usually, large superimposed text is used and the narrative voice is completely stopped. These can be thought of as informative text with audio silence. An example is a slide frame in a training video.
- *Demonstration linkage sections* are used to show demonstrations or examples of the subject under instruction. For example, in a safety video presenting the fire safety issue, there is a segment in which the narration is completely stopped and a sequence of pictures of burning houses is shown. These scenes obviously do not give any instruction or convey any concepts in a direct way, but create a 'mood' that helps the film be more realistic in presenting fire safety issues. Another example is a web page shown during a lecture in a classroom lecture video.

Discussion sections encompasses all conversations between the instructor and students, other dialogs, or multi-person discussion sections in the video. Examples are Q&A, interview, and panel sessions shown in instructional videos.

Figure 2 shows how different cues in instructional video can be analyzed to determine and interpret these narrative elements. In the sections following, we will focus on techniques to analyze the audio and video information in instructional video to automatically determine some examples of these narrative structures, namely the discussion sections, slides, web pages, and instructor segments.

3.2 SVM-Based Instructional Audio Analysis

Audio based video analysis has been an active research area for quite some time, and various audio features and classification schemes have been proposed. For instance, Kimber and Wilcox applied hidden Markov models (HMMs) to classify audio signals into music or silence or speech using cepstral

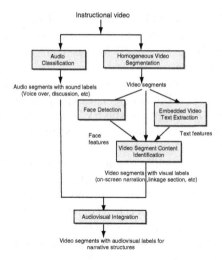

Fig. 2. Audiovisual analysis for narrative structures in instructional video.

features [185]. In [280], Saunders separated broadcast speech from music by thresholding zero-crossing rate (ZCR) and energy contour features. To find an optimal classification scheme, Scheirer and Slaney [284] examined four different classification frameworks for speech and music discrimination.

More recently, researchers have started to take additional sound types into consideration. For instance, Srinivasan et al. classified audio signals into speech, music, silence or unclassified sound type using fixed thresholds [309]. Zhang and Kuo also used a similar thresholding scheme to distinguish eight different audio types including silence, speech, music, song, environmental sound and their various combinations [370].

Although acceptable results have been obtained with these schemes, most have either considered too few audio types, or applied a threshold-based classification scheme, which makes them either less applicable or less robust to analyzing more complex audio content. To overcome these two drawbacks, we apply the Support Vector Machine (SVM) technology to the audio classification task. Lu et al. also employed SVMs in their work [214], which hierarchically classified audio signals into five classes. Particularly, it first distinguished silence from non-silence, then non-silence signals were classified into speech or non-speech. Next, non-speech segments were further classified into music and background sound, while speech segments were classified into pure speech and non-pure speech. Promising results have been reported, yet this hierarchical classification scheme presents two drawbacks: (i) if the signal is misclassified in an earlier stage, it will never reach the correct final type; (ii) it does not distinguish speech with noise from speech with music, which are two important audio classes on their own right in instructional video.

Seven audio classes are considered in this chapter which are pertinent to instructional videos including corporate education videos and professionally produced training videos. They are *speech, silence, music, environmental sound, speech with music, speech with environmental sound,* and *environmental sound with music.* Four binary SVM classifiers are trained to recognize these audio types in parallel.

A Support Vector Machine (SVM) is a supervised binary classifier which constructs a linear decision boundary or a hyperplane to optimally separate two classes [41]. Since its inception, the SVM has gained wide attention due to its excellent performance on many real-world problems. It is also reported that SVMs can achieve a generalization performance that is greater than or equal to other classifiers, while requiring significantly less training data to achieve such an outcome [337]. So far, SVMs have been applied to various tasks such as image and video content analysis and annotation, handwritten digit recognition, text classification, speech recognition and speaker identification [337]. However, SVMs have not been well explored in the domain of audio classification, and this is actually one of the major reasons that motivated our investigation of this technique.

3.2.1 Audio Feature Extraction

To classify the audio track of an instructional video, we first uniformly segment it into non-overlapping 1-second long clips, then various features are extracted from each clip to represent it. Currently, 26 audio features are considered in this work, which are chosen due to their effectiveness in capturing the temporal and spectral structures of different audio classes. A brief description of these features are given below. Readers should refer to [309, 370] for more detailed discussions.

- Mean and variance of short-time energy (STE). The energy is computed for every 20-ms audio frame which advances for every 10-ms.
- Low ST-energy ratio (LSTER). LSTER is defined as the ratio of the number of frames whose STE values are less than 0.5 times of the average STE to the total number of frames in a 1-second clip.
- Mean, variance and range of short-time zero-crossing rate (ZCR). ZCR is also computed for every 20-ms frame which coarsely measures a signal's frequency content.
- High ZCR ratio (HZCRR). HZCRR is defined as the ratio of the number of frames whose ZCR is above 1.5 fold average ZCR rate to the total number of frames in a 1-second clip.
- Mean of the spectrum flux (SF). SF is defined as the average variation of the spectrum between adjacent two frames in a 1-second clip.
- Mean and variance of energies in four frequency subbands. With 11KHZ sampling rate, we define the four frequency subbands to be [0, 700HZ], [700-1400HZ], [1400-2800HZ], and [2800-5512HZ].
- Mean and variance of energy ratios of the above four frequency subbands. The energy ratio of subband i is the ratio of its energy to the sum of the four subband energies.

- Harmonic degree (HD). HD is the ratio of the number of frames that have harmonic peaks to the total number of frames. Fundamental frequency is computed for measuring the signal's harmonic feature.
- Music component ratio (MCR). MCR is determined from the signal's spectral peak tracks which remain at the same frequency level and last for a certain period of time for most of musical sounds [370].

3.2.2 Audio Classification Using Combinations of SVMs

Every 1-second clip is classified into one of seven audio classes which include four pure audio classes: *speech, silence, music, environmental sound*, and three sound combinations: *speech with music, speech with environmental sound*, and *environmental sound with music*. Four binary SVM classifiers are trained for this purpose, which discriminate between speech and non-speech (*spSVM*), silence and non-silence (*silSVM*), music and non-music (*musSVM*), and, environmental sound and non-environmental sound (*envSVM*). A decision value DV is output from each classifier for every test clip, whose sign determines the predicted class. For instance, if the output of the *spSVM* for clip s_i is positive, then it contains speech; otherwise it is non-speech. However, since it is a multi-class classification task, additional decision rules are constructed based on an analysis of four DV values obtained from the four SVM classifiers for clip s_i.

We smooth the classification results by removing isolated audio types since a continuous audio stream does not have abruptly and frequently changed audio content. We also group temporally adjoining 1-second clips together if they share the same sound type. As a result, the entire audio stream will be partitioned into homogeneous segments with each having a distinct audio class label.

3.2.3 Experimental Results

We used the SVM^{light} software package [172] for SVM training and testing. The training data test included 40-mins of speech, 15-mins of environmental sound, 7-mins of music and 7-mins of silence which were collected from various corporate education and training videos. All data were sampled at 11KHZ rate, with mono channel and 16 bits per sample.

To find the optimal SVM parameters such as *kernels, variance, margin* and *cost factor*, we also hand-labelled approximately 9-mins of speech, 2-mins of silence, 3-mins of music and 1-min of environmental sound as validation data. Based on the validation results, we chose the radial basis function (RBF) as the kernel and set parameters γ to 5 and C to 10. Approximately a classification accuracy of 99.1% and 98.6% was achieved on the training and validation data, respectively. The entire training process took approximately 2 minutes.

The test set was collected from three education and four training videos which amounted to 185 minutes in total. Various types of sounds such as

Table 1. Overall classification accuracy with seven test videos.

Video	Length (min)	Classif. Accuracy
Train1	15:13	95%
Train2	19:19	98%
Train3	21:28	96%
Train4	13:23	99.5%
Education1	20:12	98.9%
Education2	40:12	99%
Education3	55:50	99%

background noise, speech over background noise and speech over music were contained in these videos. There was no overlap between the test and training sets. The overall classification accuracies for the seven test videos are reported in Table 1. Good classification performances have been observed, especially for "Train2", "Train4" and educational videos, which have clean speech and less noisy audio background.

3.3 Narrative Constructs in Instructional Media: Discussion Sections

By *discussion scenes*, we refer to those segments where students interact with their instructor such as asking questions and making comments, as opposed to narration scenes where the instructor gives fairly uninterrupted lectures. Since in a typical e-classroom setup, there are no cameras dedicated to capture the activity of the classroom audience, a discussion scene usually does not reveal itself with visual cues. In this difficult context, we detect classroom discussion scenes by purely using the accompanying audio cues.

The key issue in locating discussion scenes is to correctly detect speaker change points as they form the most informative audio cues. Many approaches have been proposed so far to tackle similar or related problems such as speaker tracking, speaker turn detection and speaker segregation. A general solution proposed in these approaches is to first compute distances between adjacent speech segments, then denote the points where local maxima occur as speaker change candidates. Various distance measures such as Bayesian Information Criterion (BIC) [53], Generalized Likelihood Ratio (GLR) [31], Kullback-Leibler (K-L) distance [215] and Vector Quantization (VQ) distortion [228] have been proposed for segment comparison purposes. Generally speaking, these distance-based approaches are algorithmically straightforward and easy to implement, yet they are very sensitive to background noise. To achieve more robust results, some work has applied Hidden Markov Models (HMM), Gaussian Mixture Models (GMM), or VQ codebook to represent and identify different speakers. For instance, Lu *et al.* [215] proposed to model speakers using incrementally updated quasi-GMMs and locate speaker change boundaries based on the BIC criterion. The adaptive GMM or HMM speaker modelling has also found its applications in speech recognition [371].

Since we have a single instructor in each educational video typically, we model the instructor with an adaptive GMM which is incrementally updated as the audio stream proceeds. The speaker change points are then located by comparing successive speech segments with this model using the K-L distance metric. Meanwhile, a four-state transition machine is designed to extract discussion scenes in real-time. Overall, the system contains the following three major modules: 1) audio content pre-processing which includes audio classification, speech-silence segmentation, and linkage phrase detection; 2) instructor modelling; and 3) discussion scene extraction. Each module will be discussed in the following sections.

3.3.1 Audio Preprocessing

Given the audio track of an instructional video, we first classify each 1-second audio segment using the SVM-based technique described before into one of the following sound types including speech, silence, music, environmental sound and their combinations. Then adjacent segments with the same sound type are merged to form homogeneous audio clips.

Next, for each clip that contains speech signal, we further separate pure speech segments from the background silence (*e.g.*, short pauses) using the K-means clustering approach as described in [204]. We then detect and remove linkage phrases from the obtained speech segments. By linkage phrase, we mean those spontaneous speech patterns such as "um", "ah" and "yeah" which frequently occurs in instructional videos. Although linkage phrases are part of speech, they are acoustically different from regular speech signals. The waveforms of a linkage phrase "um" and a regular speech signal show that the linkage phrase has a more stable energy envelope. Most of its energies is constrained within a narrow frequency band, while those of the regular speech signal are more sparsely distributed. Since the existence of linkage phrases can severely affect the subsequent speaker change analysis, it is necessary to remove them at this stage.

Two audio features are extracted for this purpose: the zero-crossing rate (ZCR) and the audio spectrum centroid (ASC) that identifies the frequency band with the highest signal energy. A threshold-based scheme is able to separate the linkage phrases from regular speech.

3.3.2 Modeling Discussion Scenes

The speech segments thus processed, are analyzed to determine the presence of discussions. We first represent the speech segments in an appropriate feature space, then a GMM model is initialized for the instructor. Next, we employ a four-state transition machine to detect discussion scenes in real-time while updating the GMM model on the fly.

Line spectrum pairs (LSP) is a type of spectral representation of linear predictive coefficients (LPC), which can effectively capture audio spectral characteristics under noisy environment [215]. In our work, we represent each speech segment with the mean and covariance of 14-D LSP feature vectors calculated from its underlying audio frames (30ms long). The K-L distance metric is then utilized to measure the LSP dissimilarity between two speech segments S_1 and S_2. Each speech segment S_i is represented as (μ_i, Σ_i), where Σ_i is the estimated LSP covariance matrix and μ_i is the estimated mean vector of the segment.

Assuming that the instructor initiates the lecture (this can be changed as discussed later), we use the first speech segment S_1 (μ_1, Σ_1) to initialize his/her GMM model, i.e., to form its first mixture component. Then, given each subsequent segment S_i (μ_i, Σ_i), we compute its distance from the model and determine whether it is from the instructor. If yes, we proceed to update the model as follows.

Denote $C_0 = \arg\min_{1 \leq j \leq C} \{dist(S_i, C_j)\}$, where C_j is the jth mixture component and C is the total number of components in the current model. Distance $dist(S_i, C_j)$ from S_i to C_j is calculated. Now, if $dist(S_i, C_0)$ is less than some threshold T_1, i.e., if S_i is acoustically similar to C_0, we update component C_0 (μ_0, Σ_0) to accommodate S_i's acoustic characteristics.

Otherwise, if $dist(S_i, C_0) \geq T_1$, we then initialize a new mixture component to accommodate S_i by setting its mean and covariance to those of S_i's. Note that once the total number of mixture components C reaches an upper limit (currently set to 32 based on experimental results), no more components will be added to the model. Instead, we replace the oldest component with the newly generated one. Finally, we update the weights w_j for all components C_j $(j = 1, ..., C)$ accordingly.

By generating and updating the model in this fashion, we are able to accommodate the instructor's voice variations over time, thus capturing his/her acoustic characteristics better. Although the GMM generated using our approach is very different from the EM (Expectation Maximization)-trained GMM, nor is it the same as the ones proposed in [215, 371], it can be obtained in real-time with reasonable performance as shown by our experimental results.

3.3.3 Discussion Scene State Machine

A four-state transition machine is designed to extract discussion scenes at this stage. As shown in Fig. 3, it has the following four states: *discussion, narration, leaving-discussion* and *leaving-narration*, which are determined from our analysis of a lot of educational videos. The machine starts from the *narration* state. Now, denoting the next incoming speech segment by S_i whose distance from the instructor's model equals $D = \sum_{j=1}^{C} w_j \times dist(S_i, C_j)$, we determine the state transition flow as follows.

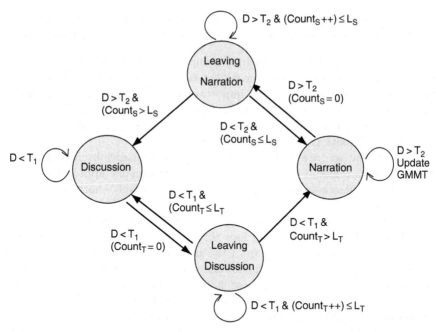

Fig. 3. A four-state transition machine for discussion scene extraction in instructional video analysis.

1. *The current state is narration.* In this case, if D is less than some threshold T_2, *i.e.,* segment S_i is from the instructor, we remain in the same state and update the instructor's GMM model. Otherwise, we mark it as an *instructor-student change* point, and the machine leaves the narration state.

2. *The current state is leaving-narration.* This is a transient state, mainly used to accommodate short cut-ins such as a quick question or comment from the students during the lecture. As shown in the figure, if $Count_S$, which counts the number of students' speech segments, exceeds threshold L_S, the machine transits to the discussion state. Otherwise, it will move back to the narration state and ignore this short discussion moment. L_S is currently set to be 5, which is applied to also avoid false alarms caused by background noise.

3. *The current state is discussion.* In this case, if D is larger than T_1, *i.e.,* segment S_i is from a student, we remain in the current state; otherwise, we declare a *student-instructor change* point and leave the discussion state.

4. *The current state is leaving-discussion.* This transient state is used to accommodate some short speeches from the instructor during a discussion. As shown, if $Count_T$ which counts the number of segments from the instructor exceeds threshold L_T, the machine transits to the narration

state; otherwise, it goes back to the discussion state and accommodates the instructor's speeches as part of discussion. L_T is currently set to be 3.

To summarize, this state machine groups a series of continuous speeches from students as a discussion scene while tolerating short instructor cut-ins at the same time. T_1 and T_2 are two important thresholds used in GMM model update and discussion scene detection processes, and should be automatically and intelligently determined. In our work, we derive their values from the statistical distribution of the K-L distance.

As a comparison, we examined the system performance by modelling the instructor with a global GMM model. This is a reasonable approach since an instructional video is usually captured within a relatively short period (*e.g.,* a few hours).

To automate the data transcribing process for global model training, we first classify all speech segments into three groups with group 1 containing speech from students, group 3 from the instructor and group 2, the mixed speech. Two features, namely, the loudness mean and variance are extracted for this purpose, which are selected based on the observation that the instructor's voice is normally louder than those of the students since he uses a microphone. The EM algorithm is then used to train the model using speeches from group 3. To be consistent with the adaptive GMM approach, the global model also contains 32 mixture components. Finally, the same state machine is applied to extract discussion scenes except that no more model update is needed and thresholds T_1 and T_2 need to be adjusted accordingly.

Currently, we use this global model to also determine the starting state of the transition diagram, as well as to locate the instructor's first speech segment for initializing his/her GMM model.

3.3.4 Experimental Results

We tested our discussion scene (DS) detection scheme on five instructional videos recorded from the IBM MicroMBA educational program. Lengths of these videos vary from 60 to 90 minutes. In the test set, MBA I and II are lectured by the same instructor to whom we refer as A, MBA III and IV by instructor B, and MBA V by instructor C. Six statistical ratios namely, *DS-Precision*, *DS-Recall*, *SW-Precision*, *SW-Recall*, *SWDS-Precision*, and *SWDS-Recall*, are computed to evaluate the detection results. The first two evaluate the precision and recall rates at the scene level, while the second two rates aim to measure the localization (boundary) accuracy of those detected DSs. Finally, we fuse them into the last two rates for an overall system performance evaluation as: SWDS-Precision $= w_1 \times$ DS-Precision $+ w_2 \times$ SW-Precision, and SWDS-Recall $= w_1 \times$ DS-Recall $+ w_2 \times$ SW-Recall. w_1 and w_2 are two weights that sum up to 1. Currently, we set w_1 to 0.6 and w_2 to 0.4, which reflects the fact that to correctly detect DSs is more important than to precisely locate their boundaries.

Figure 4 shows the sample waveform and the discussion scene detected.

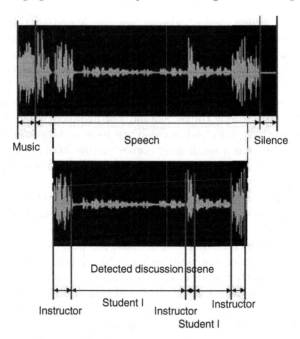

Fig. 4. Sample audio track and discussion section from an instructional video.

Overall, we found that the adaptive approach has achieved an average 88% and 93% SWDS precision and recall rates, which has well outperformed the global model approach with 72% and 83% rates. Nevertheless, the performance of the global method could have been slightly improved if more training data was used. We also observed that the system tended to have more false alarms than misses which were mainly caused by background noise, overlapped and distorted speeches (*e.g.,* speech with laughter).

In addition, the performance of the global approach is consistent with that of the adaptive approach, *i.e.,* whenever the adaptive approach performs well on certain videos, so does the global approach. This is reasonable as they are modelled from the same speeches except that one is adaptively updated while the other is fixed.

3.4 Visual Narrative Constructs in Instructional Media

An instructional video can also be partitioned into homogeneous segments where each segment contains frames of the same image type such as slide or web-page; then we can categorize the frames within each segment into one of the following four classes: *slide, web-page, instructor* and *picture-in-picture,* by analyzing various visual and text features in the frames. By classifying video frames into semantically useful visual categories, we are able to better understand and annotate the learning media content.

Four distinct frame content classes are considered in this work which include *slide, web-page, instructor* and *picture-in-picture*. By definition, the instructor frame contains a close-up to mid-range shot of the instructor or speaker in the video, while a slide or web-page frame contains a close-up view of a slide (or a PowerPoint presentation page) or a web-page. In contrast, a picture-in-picture frame has an embedded sub-image which is usually independent of the surrounding frame content. For instance, when the instructor launches a demo that plays back a video, we have such type of frames. Note that, an image that has a small inset corner picture of the instructor that is usually seen in e-seminars, is not considered to be of this type. Figure 5 shows four typical frames of these content types.

Video content analysis has been studied for decades, yet very few research efforts have reported on the frame classification topic. The work that we are aware of was from Haubold and Kender [138], where they classified keyframes into six categories which include board, class, computer, illustration, podium and sheet, using a decision tree based scheme. Nevertheless, they use too many heuristics specific to their data in their approach which makes it impractical for classifying generic learning videos. For instance, they assume that computer frames are all black-bordered and contain horizontal lines, while board and podium frames have a dominant green color.

Our work presents a hierarchical video partition and frame classification scheme that contains the following four modules: (1) homogeneous video segmentation which partitions a video sequence into a series of cascaded segments. Each segment contains frames of the same image type; (2) picture-in-picture segment identification which exploits its distinct visual characteristics in content variation and variation areas for identification; (3) instructor segment identification which utilizes extracted face and image tone information to recognize the instructor frames; and (4) slide and web-page discrimination which recognizes slides based on extracted text features. Each module is detailed in the following sections.

3.4.1 Homogeneous Video Segmentation

In this module, we measure the content change between two neighboring frames in a video using a *peak-based* frame comparison approach which is able to tolerate minor content changes such as those caused by digitization noise, jerky camera motion and illumination changes, to a certain degree.

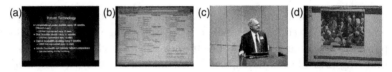

Fig. 5. Typical frames that contain (a) slide, (b) web page, (c) instructor, and (d) picture in picture.

Given a frame, we first partition it into $MD \times ND$ grid (MD and ND are currently set to 8). Then, for each grid cell, we calculate its B-bin color intensity histogram (B is currently set to 64) and locate all distinct peaks. Next, we associate each peak with its component bins (*i.e.*, the bins that form the peak), and term it as a *peak category*.

For each frame pair (say, frames j and $j+1$), we compare the content of their i^{th} cells ($i = 1, \ldots, MD * ND$) by examining the overlap of their peak categories. Specifically, denote these two cells by $C_{j,i}$ and $C_{j+1,i}$, and assume that $C_{j,i}$ has a set of peak categories $PS_1 = \{PC_1, \ldots, PC_K\}$, and $C_{j+1,i}$ has $PS_2 = \{PC'_1, \ldots, PC'_K\}$, then for any m ($1 \leq m \leq K$), if PC_m and PC'_m share a majority of their histogram bins, *i.e.*, they present similar color distributions, we claim that $C_{j+1,i}$ and $C_{j+1,i}$ have same content. Note that if the numbers of peak categories disagree in these two cells, we immediately reject them without further processing. By only concentrating on the major peaks that capture dominant cell colors, and by tolerating minor variations in bin populations, we are able to better reveal the true content changes. We refer to the cell whose content remains the same for two consecutive frames as *background* (BG) cell, otherwise, it is a *foreground* (FG) cell.

With this computation, the difference between two frames is quantified by the number of their foreground cells. A larger number indicates a distinct frame difference. The content change boundaries are then located from the FG cell distribution based on the identification of distinct peaks. We locate those peaks that serve as the video segment boundaries using a *bi-directional background cell tracing* algorithm [203]. Each of these segments only contains frames of the same type that we consequently it a *homogeneous segment*. Note that here, we do not require that two adjacent segments display different types of frames, but only that each segment displays frames of the same type. Therefore, a segmentation that has extra segments is acceptable.

3.4.2 Detecting Picture-in-Picture Segments

This module identifies segments that contain picture-in-picture (PiP) frames. Two facts are exploited in identification: (i) a PiP segment usually presents a much larger content variation since the sub-image content keeps changing over time; (ii) this content change is confined to a local area (*i.e.*, the sub-picture area) in the frame.

To measure the content variation with a homogeneous segment S, we first calculate the number of foreground cells over every frame pair, then use their variance as the index. Intuitively, the larger the variance, the higher the content variation. Meanwhile, the image locality feature is examined by checking if all cells whose contents changed over S are bounded to a confined image area. Once all the PiP segments are detected, they are excluded from the subsequent processes.

3.4.3 Detecting Instructor Segments

To identify frames that contain a close-up view of the instructor, one straight-forward approach is to detect human faces. Nevertheless, as the instructor generally keeps moving in front of the podium, it is not always possible to locate his or her face reliably with an automatic face detector. Therefore, one more source of information, namely, the dominant image tone, has been integrated into the analysis to help improve the detection accuracy.

Given a non-PiP segment, we extract the following types of face information: face-contained frame ratio fr, and duration of the longest face subsequence fd. Specifically, $fr = \frac{\sum_{i=1}^{N} HasFace(f_i)}{N}$, where $HasFace()$ is a binary face function which equals 1 when a face is detected in frame f_i. N is the total number of frames in the segment. An instructor segment should have a higher fr ratio than a segment containing slides or web-pages. Feature fd denotes the duration of the longest subsequence s within the segment, where every frame of s contains a detected face. This feature comes from the observation that when a face is truly detected in a frame, it should remain detectable in a few subsequent frames due to the continuity of video content. A face may be falsely located in a slide frame, yet the possibility of making the same mistake in a row will be much lower.

Figure 6 (a) shows the statistics of the two features (normalized) for a test video. The feature curves are consistent with the rationale well, *i.e.,* if both the features reach their local maxima over a segment, it will be truly an instructor segment. Notice that segment 9, which contains web-pages, has a relatively large fd value. This is caused by faces present in web-pages. Nevertheless, its face-contained frame ratio is still fairly low. Moreover, the image tone information, presented in the next section, could be further applied to avoid false detection.

3.4.4 Identifying Candidate Slide and Web Page Segments

Generally speaking, when the instructor uses slides to assist his or her talk, 80% of the video frames will be slides. Consequently, we can estimate the

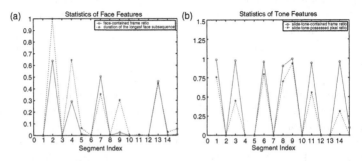

Fig. 6. Statistics of the (a) face features and (b) tone features for a test video.

major slide tones by identifying the dominant video tones. On the other hand, an instructor frame should normally present rather different image tones from those of slides. Relying on this fact, we extract the following two tone features from each segment: ratio of the frames that contain slide tones (sr), and ratio of the pixels that possess slide tones (sd).

We derive the slide tones by identifying the dominant video tones, which in turn, are acquired from the dominant image tones in each frame. Specifically, given a frame, we first compute its B-bin color intensity histogram, then we sort all bins based on their populations in a descending order. Next, we identify the bins that jointly cover $p\%$ of total image pixels and designate them as the dominant image tones. The percentage threshold p is currently set to 70. The slide tones (and also, the dominant video tones) are then set to be the tones corresponding to the highest peak in the tone histogram.

For any frame f_i in a given segment, we examine the overlap between its major image tones and the slide tones. If the overlap is high, say, more than 90%, we mark it as a slide candidate. We also sum up all pixels that possess slide tones in f_i. As a result, after we process all frames in the segment, we obtain a list of slide candidates (denote its number by N_s), and the number of slide-tone-possessed pixels (denote it by N_p). The two tone features, sr and sd, are then calculated as $sr = \frac{N_s}{N}$, and $sd = \frac{N_p}{NP}$, where N and NP are the total number of frames and pixels in the segment, respectively.

Figure 6 (b) shows the statistics of these two tone features for the same test video. One interesting thing to observe is that this figure has presented exactly the opposite features from those in Fig. 6 (a) (note that segments 5, 10 and 12 are PiP segments whose tone features are thus set to zero). Whenever the two face features reach their local maxima, the two tone features will surely hit their local minima over the same segment. This leads to the fact that the instructor segment, which has higher face-contained frame ratio and longer face subsequence, usually has non-slide tones since it mostly contains the instructor, the podium, and the classroom. We also notice that segment 9 discussed before has now presented fairly large tone values, which thus disqualifies it from being considered as an instructor segment.

The post-processing step aims to find the rest of instructor segments that may have been missed out due to a lack of reliably detected human faces. This is achieved by comparing a segment's group color histograms or major image tones with those of a confirmed instructor segment. This is adopted in the current work due to its low computation complexity.

3.4.5 Discriminating Slide and Web-Page Segments

This module discriminates slides from web-pages which are the last two image types under consideration. Since a homogeneous slide/web-page segment contains multiple slides/web-pages where each slide/web-page is displayed over an extended temporal period, we attempt to first extract distinct slides/web-pages from each segment, then categorize them into one of the two classes. We refer to these distinct slides/web-pages as segment keyframes.

Segment keyframe identification: To identify the keyframes from a slide or web-page segment, one straightforward approach is to first partition it into smaller fragments where each fragment displays the same slide; then we designate any frame from a fragment as its keyframe. We compare two frames in terms of their Canny edge maps, as described in the following four steps.

1. *Canny edge detection.* We use the standard Canny edge detection algorithm to extract the edge map [308]. The parameter σ, as well as the two thresholds (low and high) applied in the detection process are empirically determined.
2. *Canny edge histogram calculation.* After obtaining a frame f_i's edge map, we sum up the number of edge points residing in each of the $MD \times ND$ image cells, and construct an edge histogram H.
3. *Canny edge histogram comparison.* We measure the difference between two Canny edge histograms using the city-block metric.
4. *Content change boundary detection.* Finally, we locate all distinct peaks (including the one that extends over a temporal period) from the edge histogram difference map. The "peak-tracking" approach is again used for this purpose. Each segment is thus partitioned into a series of cascaded *fragment* where each fragment displays the same slide or web-page.

Once we obtain the fragments, we choose their middle frames as their keyframes, since the starting or ending frames may have been affected by a gradual content transition.

Processing steps for slide and web-page detection: This last module differentiates slides from web-pages, which is achieved by exploiting facts such as: (i) a slide generally has a distinct title line; (ii) the title line is usually well separated from other text regions in the frame; and (iii) the title line is of a relatively larger font. No such general conclusions could be drawn for web-pages, since different pages can have different layout styles. Below are the six major image processing steps in extracting two text features used to discriminate slides from web-pages.

Step 1: Image smoothing, sharpening and unwarping. A projection screen is physically built to be uniformly white, yet due to different amount of incident light on pixels, color variations result even in regions with uniform colors. Therefore, we choose the image smoothing and sharpening as the very first processing step, using the edge strength-guided smoothing [308] algorithm that has good performance in both edge preserving and noise removal. To facilitate the edge detection in subsequent processing, this step also applies image sharpening to enhance object edges [308].

Sometimes, the video sequence contains a foreshortened version of the image. This should be corrected since it affects the subsequent text feature extraction step. Since we have no access to the class/conference room setup, we have to derive the transform that relates consecutive images from the video sequence itself. We first identify a frame which contains a quadrilateral q, then we locate its corresponding rectangle r in the unwarped image. Next, we derive the affine transform that maps q to r, and use this to unwarp all other frames in the video sequence.

Step 2: Canny edge detection. This step performs a Canny edge detection to obtain a Canny edge map of the smoothed, sharpened, and unwarped image, I. Then it removes all horizontal and vertical straight lines from the map as we are only interested in text regions. Figure 7 (b) shows the edge map of a web-page in Fig. 7 (a).

Step 3: Text region extraction. This step locates the image's text regions based on the exploitation of the following two facts: 1) a text character has strong vertical and horizontal edges; and 2) text characters are always grouped in words, sentences and paragraphs. The two steps involved in this process are described below.

1. For each pixel p in the Canny edge map, we sum up the number of edge pixels in its $W \times H$ neighborhood area and set this as p's new value. We term the resulting image as I's *neighborhood edge map* (NEM). Pixels that reside in a tightly clustered text regions of I present larger values in the NEM map. The window size, which is empirically determined from the character size in test slides, is set to be 5×9. As an example, we show the NEM map of a test web-page in Fig. 7 (c).
2. We then binarize the NEM map, perform necessary morphological operations to fill the holes within text lines, and remove noisy data to obtain the final text map. The threshold applied to binarize the NEM map is currently fixed to 8. The final text map for web-page is shown in Fig. 7 (d).

Step 5: Title line detection. This step locates the image title line which: 1) is the first text line in the image; 2) is well separated from other text; and 3) has all of its text tightly grouped together. A line profile [93] is constructed, by summing up the number of text pixels in each image row. A sample line profile is shown in Fig. 7 (e). The line profile analysis is carried out in the following two steps.

1. *Peak detection.* We detect all distinct peaks from the line profile by enforcing the following two rules: 1) each peak should be distinctively higher than its neighbor(s); and 2) it should be well separated from the others.
2. *Title line detection.* Starting from the first peak, we verify its qualification to be a title line. Specifically, for the image row that corresponds to the peak tip, we calculate the average distance between every two adjacent text pixels. A large value indicates that text regions are sparsely distributed in the line, thus decreasing its possibility of being a title line. In that case, we skip the current peak, and check the next.
   In the case of Fig. 7 (a), the title line is located from the second peak as the first one is unqualified. Figure 7 (f) shows the final text map where the text corresponding to the first peak in the profile is removed.

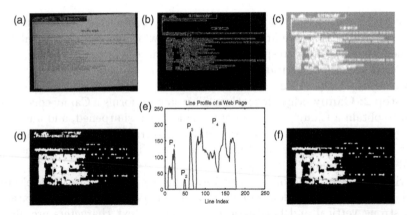

Fig. 7. (a) A web-page image after unwarping, (b) the Canny edge map after the removal of horizontal and vertical lines, (c) the neighborhood edge map, (d) the text map, (e) the line profile, and (f) the final text map after the removal of the first non-title line.

Step 6: Text feature extraction. We extract the following two text features from each frame: 1) the width of the title line that reveals the text size; and 2) the distance from the top frame edge to the title line that reveals the title position in a vertical direction. These two features could be derived from the image line profile, where feature 1 corresponds to the title peak's width, while feature 2 is the distance from its starting point to the origin.

We differentiate slides from web-pages using the k-means clustering approach.

1. For all frames which are either slides or web-pages, we first cluster them into three groups using feature 1, then re-cluster them into another three groups using feature 2.
2. From each clustering, we identify the mostly populated group g_i ($i = \{1, 2\}$) and declare that this group has a higher possibility of containing slides than the other groups. This is derived from the fact that, in a presentation-centric talk, slides usually outnumber web-pages. When a frame belongs to both g_1 and g_2, we identify it as a slide; otherwise, it is a web-page.
3. Finally, a post-processing is carried out which incorporates the knowledge of homogeneous segment boundary. If we detect that more than half of the frames in a segment are slides or web-pages, then very likely it is a slide or web-page segment. Consequently, all frames within the segment should be of the same type.

Note that, to speed up the processing, it may not be necessary to compute the text map for every frame since the frame content is fairly constant within a slide or web-page fragment. One good solution is to only compute the text map for the keyframe of each fragment, then assign this text map to its other frames.

Table 2. Performance evaluation on two test videos.

Homogeneous video segmentation					
Test Video	Hits	Misses	False Alarms	Precision	Recall
Video 1	15	1	1	93.8%	93.8%
Video 2	78	2	2	97.5%	97.5%
Slide/web-page change detection					
Video 1	40	0	2	95.2%	100%
Video 2	53	1	3	94.6%	98.1%

3.4.6 Experimental Results

We carried out some preliminary experiments to validate the algorithms presented. Specifically, two video sequences, both of which are taped from presentations, were tested. Each video was 90 minutes long. Some results are tabulated in Table 2. Note that in evaluating the homogenous video segmentation performance, we call it a false alarm when a detected video segment contains more than one type of video frame.

With video 1, two neighboring segments, which contain web-pages and instructor respectively, were mistakenly returned as one homogeneous segment. This consequently results in one miss and one false alarm. Two reasons for this error are: 1) the content transition from the web-page to instructor frame is extremely slow which results in a very small inter-frame difference; 2) the web-page forms a large portion of the instructor's frame background.

All segments that contain picture-in-picture and the instructor have been correctly identified. Moreover, good performance is also observed on the slide and web-page segment fragmentation. Finally, for the slide and web-page discrimination, without performing the post-processing, we achieved 70% and 96% classification accuracies on three slide and four web-page segments, respectively. However, after we take the homogeneous segment boundary information into account, these two rates improve to 87% and 100%, respectively. We did not achieve a perfect classification on the slides since the first two slides did not present the features as we expected.

Compared to video 1, video 2 is better edited and contains 80 homogeneous segments in total. Among them, there are 28 instructor segments, 22 audience segments which are long shots of the auditorium or rather, the audience, 22 slide segments and 8 web-page segments. Two false alarms and two misses occurred as we failed to identify the boundaries between two slide segments and their adjacent web-page segments. The slow content transition is the major reason for causing such errors. The video did not contain picture-in-picture segments and all instructor segments have been correctly recognized.

The performance of slide and web-page segment fragmentation is comparable to that of video 1, although both rates have been slightly dropped. Finally, 92% and 94% classification accuracies are achieved on the slide and web-page categorization, respectively.

4 Conclusion

By automatically classifying audio and video content into meaningful lasses, partitioning videos into homogeneous segments, and annotating them with audiovisual labels, we can build a table-of-contents for navigating instructional videos according to narrative structure. The analysis results, which can be used to populate certain metadata fields in the SCORM LOM, will facilitate the indexing, access, browsing and searching of content in e-learning.

Effective content management in e-learning requires an enterprise-wide approach to facilitate content acquisition and archival; workflow; version control; meta-tagging; file conversions and transformations; security, permissions, and rights management; and integration with multiple repositories, applications, and delivery networks.

Today, special-purpose media subsystems using proprietary technology are often built as weakly linked add-ons to applications, rather than well-integrated components in a company's IT architecture. This makes it difficult to integrate new rich media content models with existing legacy business data. It complicates the workflow in billing and commerce systems, customer service applications, and network infrastructures. Often, the add-on nature and use of proprietary technologies results in short-lived complex custom solutions. Finally, media data today is typically owned by individual applications such as news archive management, rather then being treated as common business assets. Unfortunately, this approach leads to isolated point solutions.

The full integration of digital media as a commonly available data type for business applications is critical and will cause a major transition in computing infrastructure and solution architectures.

Software providers should embed support for digital media directly into their middleware and infrastructure products to support the creation of complete solutions. This support must address the scalability, complexity, and other unique requirements associated with large amounts of unstructured data in various formats. It should provide integration points necessary for seamless implementations of component products for solutions that manage the entire content lifecycle. Development and acceptance of media standards will be especially important to enable interoperability of both media and content, minimize technology investments, streamline operations, reuse and repurpose digital media, and deliver it securely to the widest audience possible. The business needs for media are concrete; the problems we need to address are real. Our community needs to find real solutions too, for high-level semantic analysis, provisioning and management of mixed-media information, and distribution and delivery of media data to satisfy requirements dictated by business scenarios.

Acknowledgements

I would like to thank all my collaborators around the world, with whom I had the opportunity to investigate the application of the Computational Media Aesthetics approach to various video domains. I specially thank Martin, Dinh, Svetha, and Ying for their valuable contributions to our joint work over the years and to this chapter.

Part IV

Interfaces

Interactive Searching and Browsing of Video Archives: Using Text and Using Image Matching

Alan F. Smeaton, Cathal Gurrin, and Hyowon Lee

Centre for Digital Video Processing and Adaptive Information Cluster,
Dublin City University, Glasnevin, Dublin 9, IRELAND.
Alan.Smeaton@dcu.ie

1 Introduction

Over the last number of decades much research work has been done in the general area of video and audio analysis. Initially the applications driving this included capturing video in digital form and then being able to store, transmit and render it, which involved a large effort to develop compression and encoding standards. The technology needed to do all this is now easily available and cheap, with applications of digital video processing now commonplace, ranging from CCTV (Closed Circuit TV) for security, to home capture of broadcast TV on home DVRs for personal viewing.

One consequence of the development in technology for creating, storing and distributing digital video is that there has been a huge increase in the volume of digital video, and this in turn has created a need for techniques to allow effective *management* of this video, and by that we mean content management. In the BBC, for example, the archives department receives approximately 500,000 queries per year and has over 350,000 hours of content in its library[1]. Having huge archives of video information is hardly any benefit if we have no effective means of being able to locate video clips which are of relevance to whatever our information needs may be.

In this chapter we report our work on developing two specific retrieval and browsing tools for digital video information. Both of these are based on an analysis of the captured video for the purpose of automatically structuring into shots or higher level semantic units like TV news stories. Some also include analysis of the video for the automatic detection of features such as the presence or absence of faces. Both include some elements of searching,

---

[1] Evans, J. The future of video indexing in BBC, at TRECVid Workshop 2003, Gaithersburg, MD, 18 November 2003.

where a user specifies a query or information need, and browsing, where a user is allowed to browse through sets of retrieved video shots. We support the presentation of these tools with illustrations of actual video retrieval systems developed and working on hundreds of hours of video content.

2 Managing Video Archives

The techniques we use in the tools described in this chapter represent just some of the available approaches to managing digital video information. In the first instance, the easiest, and most useful way to organise video archives is to use raw metadata, created at the time the video is created, to index and provide subsequent access to the video. In the case of CCTV, for example, to access video at a given point, security personnel employ a combination of which camera, and what date and time, and this is usually sufficient to allow users to retrieve the video clips they are looking for. If a more refined or accurate content-based search is required then the raw metadata will not be enough and many archive libraries will annotate video content by hand. This can take up to 8 or 10 times real-time (i.e. 8 to 10 hours to hand-annotate 1 hour of original video) and is thus clearly very expensive but is used extensively in TV archives worldwide. To ensure some consistency across annotators and across time, they typically each use an ontology of only some thousands of terms which creates a structured relationship among the pre-defined set of index terms. Figure 1 shows an example of such an annotation system, used as part of TRECVid (see Sect. 4.1) for annotating broadcast news video [211], where the video shot currently being annotated is being assigned the "tags" *standing, outdoors, trees, greenery, water body, waterfall, microphone, female speech and female face*. These annotations can subsequently be used in searching or browsing.

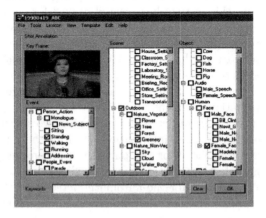

Fig. 1. Sample Manual Annotation of Video

Manual annotation is, naturally, very expensive and used only where there is a specialist need for high-quality searching, but it is not scalable to huge video archives. In the case of a user accessing an archive of broadcast TV or movies on a home DVR to find the exact clip in the movie "Minority Report" where Tom Cruise uses a video browsing system with a gesture-based interface, then manual annotation of the video content will not be available. The movie name or date/time of transmission and recording, and some clue as to how far into the movie that scene occurs will probably be enough only to locate the region of the movie where the scene occurs. The user is probably then going to need to *browse* through the video to locate the exact scene. Consider also the case of a user travelling on business and accessing an archive of tonight's TV news from their local TV station via a web interface to a video archive system. The user doesn't want to play the full 30 minutes of news but will want to *browse* through the stories, skipping those not of interest based on a story skip, possibly of video keyframes or of dialogue, and playing video clips of those stories of interest. Conventional VCR-type controls like play, pause, fast-forward and rewind can be used here, as can more intelligent approaches such as pause detection and removal and variable speed fast forward [267].

However, there is a lot more that can be done to help a user locate desired conent, for example the Físchlár-TV [299] system which is a web-based shared video retrieval system that lets users record, *browse* and playback television programmes online using their web browser. A programme recorded by one user enters a shared repository and can then be viewed by any other user of Físchlár-TV. The total video archive size is about 400 hours of video and operates as a first-in first-out queue which usually results in a programme being available for just over three weeks before being removed from the archive to make way for newly recorded programmes. TV schedules are used to allow a user to record a programme by simply clicking on a hyperlink. By default, all programmes in the archive are sorted by date and time in decreasing order of freshness and are listed in on the left side of the interface (see Fig. 2). Selecting a programme title will cause full programme details to be displayed on screen. Each recorded programme is represented by metadata (title, date, time, and a short description) and video keyframes which are extracted automatically from the programme and presented on screen for the user. In this way a user can *browse* through the content of a programme, seeking a desired section and when the desired section is found (for example, that scene in Minority Report) clicking on the keyframe begins playback of video from that point.

While a video retrieval system such as Físchlár-TV is clearly very useful, in allowing a user to quickly browse an entire TV programme by examining a collection of keyframes, the user still needs to know which programmes to browse. However, when presented with a large archive of content, a user will, in many cases, be unsure of what programme they are looking for, for example, a video archive that contains all recordings of a late night chat show, how can a user, without knowing the date of broadcast, find the interview with Madonna where she throws a glass of water in the host's face? In this case

Fig. 2. Browsing a recorded video programme in the Físchlár-TV System

there is a need for video *searching* through the actual video content, using some keywords from the dialogue between Madonna and the chat show host [296] or if CCTV video, using some representation of the object corresponding to the CCTV suspect [194].

From this introduction to interactive searching and browsing of video archives we can already see that there are at least three separate ways in which we may want to access digital video information; using raw or annotated *metadata* as a basis for searching, *browsing* through the actual video content and *searching* through the actual video content. Other content-based access tools could include summarisation, automatic gisting and highlight detection but we are not concerned with those in this chapter. Neither are we concerned here with techniques for searching through raw metadata. Instead we concentrate here on techniques for supporting interactive searching and browsing video content based on using text and image matching.

As a result of extensive research in the very recent past there are now robust, scalable and effective techniques for video analysis and video structuring which can turn unstructured video into well-formed and easy to manage video shots. There are also semi-automatic techniques for video object extraction, tracking and classification though these are not yet scalable to large video archives. There are good techniques available for recognising features in video, from simple features such as indoor/outdoor and faces/no faces present, to the more challenging naming of individual faces or naming of buildings and locations. Many of these techniques have been developed for the purpose of automatic video analysis on large video archives which can, in turn, support searching and browsing. Ideally state-of-the-art systems such as Informedia from CMU [155] or MARVEL from IBM Research [152] would be able to manage tens or hundreds of thousands of hours but as of now they are able to manage hundreds or maybe just thousands of hours of video.

The Informedia System, developed as part of ongoing research at CMU since 1994 has developed and integrated new approaches for automated video indexing, navigation, visualisation, search and retrieval from video archives. In a similar way to the two retrieval systems detailed later in this chapter, the Informedia system brings together various strands of video retrieval research into one large system that provides retrieval facilities over news and documentary broadcasts (from both TV and radio) in a one terabyte video archive. The Informedia system combines speach recognition, image feature extraction and natural language processing technologies to automatically transcribe, segment and index digital video content. Key features of the Informedia system include the extraction of name and location data from the videos, face identification, summary generation, dynamic video linkage, event characterisation and novel visualisation techniques.

MARVEL (Multimedia Analysis and Retrieval System), from IBM research is a prototype multimedia analysis and retrieval system, the aim of which is to automatically annotate multimedia (not just video) data by using machine learning techniques that model semantic concepts. MARVEL organises semantic concept ontologies and automatically assigns concept labels to video data, thereby reducing the need for human annotation of content from 100% to only 1-5% (which is required for the machine-learning processes to operate effectively). MARVEL is migrating from a current ontology of about 100 concepts to a large scale ontology of 1,000 concepts, designed to model broadcast news video. Given the ability of MARVEL to automatically determine concepts occurring in video data, a user search system incorporates this functionality along with text search functionality to produce a powerful multimedia retrieval system.

The video analysis techniques used in such video retrieval systems, and the subsequent video searching and browsing, are the focus of the work reported in this chapter, with the rest of this chapter being organised as follows. In the two sections to follow we provide system descriptions of the Físchlár-News and Físchlár-TRECVid systems developed for accessing an archive of RTÉ broadcast TV news in the case of Físchlár-News and for accessing an archive of CNN and ABC TV news in the case of Físchlár-TRECVid. In Sect. 5 we illustrate how each system supports both searching and browsing in different ways because the information needs which each was designed to address are very different. Despite the very different nature of the underlying information needs, and the resulting systems, we are able to show how the searching and browsing operations in both systems are very tightly intertwined in each system, which underscores the main point of this chapter which is to stress how equally important search and browse are for video navigation.

3 Físchlár-News: System Description

Físchlár-News is an online archive of broadcast TV news video which makes use of various content-based video indexing techniques to automatically structure TV news video to support searching, browsing and playback of the news video on a conventional web browser. An example usage scenario would be a user who is travelling and wishes to keep up with news events at home, but who does not have the time to view an entire news programme, rather would like to be able to view news stories from both missed news programmes and the entire archive.

The system's automatic processing of video is illustrated in Fig. 3. At 9 o'clock every evening, Físchlár-News records the TV news from the Irish national station RTÉ into MPEG-1 (top-left of Fig. 3), along with the closed-caption data (spoken dialogue text supplied by the broadcaster) transmitted at the same time. The encoded MPEG-1 file goes through a series of automatic content-based indexing processes, starting with *shot boundary detection* [39] which segments the news video into individual camera shots, followed by *advertisement detection* [277] which removes TV ads that sometimes appear at the beginning, the end, and in the middle of the broadcast news. The ads-removed parts of the video are then subject to *news story segmentation* [241] which involves a number of content analysis techniques including speech/music discrimination and anchorperson detection, and their output combined using a machine-learning technique to determine more reliable news story boundaries.

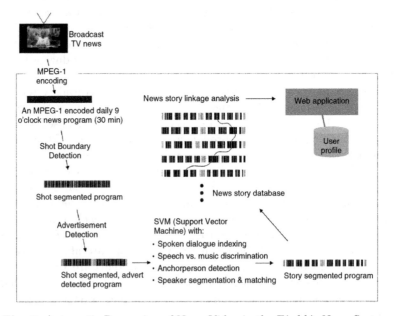

Fig. 3. Automatic Processing of News Video in the Físchlár-News System

The system requires these multiple evidence combinations at this stage because accurate news story segmentation is still a major challenge in the video retrieval community with various approaches being tried [64, 188]. The closed-caption signal is also indexed with conventional IR techniques and aligned with the corresponding video data to support text-based searching. The outcome of all this is that a day's broadcast TV news is automatically structured into topical, individual news stories each of which is again segmented into a number of shots. Once this stage is reached, the structured video is stored in the news story database in which all previous days' news stories have been indexed and are available for retrieval (top-right of Fig. 3). Currently this database contains over 3 years of news, amounting approximately 8,000 news stories. These news stories are available via a conventional web browser, allowing news story-based searching, browsing and playback.

Figure 4 shows a screen shot of the web interface. On the left side of the screen the monthly calendar allows access by date. When a user clicks on a date, that day's news stories are presented on the right side of the screen, each story with an anchorperson keyframe, and first two lines of closed-captions. In Fig. 4 a user is searching for stories related to politician Paul Bremer. The user typed in 'Bremer' in the query box and clicked on the GO button. The search term was matched against the indexed closed-captions and the resultant news stories returned. In Fig. 4 five news stories were retrieved as a result and presented on the right side of the screen. The user can simply play any of the retrieved stories by clicking on the 'Play this story' button at the end of each closed-caption summary which will pop up a video player plug-in and start playing the story, or browse more detail of the story by clicking on the title

Fig. 4. Text-based Searching and List of Stories as Search Result

Fig. 5. Browsing Shot-Level Detail of a News Story

of the story. Figure 5 shows a screen when the user selected a fourth story in
the retrieval result from Fig. 4 (story dated 23 August 2003) to browse more
detail. The user is then presented with a "storyboard" of the story, a full list
of keyframes from each camera shot contained in the story with interleaved
closed-caption text, providing detail of the story for quick browsing (right
side of Fig. 5). The shot-level storyboard shows all major visual content of
that story in one glance, without requiring a video playback, and is featured
in the majority of digital video retrieval systems available today. Clicking
on any of these keyframes will pop up the player plug-in and start playing
from that point in the story onwards, enabling playback at a particular point
within a story.

The user can also browse through other similar stories within the archive,
to trace the development of the story in last few months or to browse other
stories related in some way (e.g. the same company is involved or same person
mentioned) for more serendipitous browsing. Below the storyboard of a news
story with keyframes and closed-captions, the user can see the ten stories
that are related to that story as shown in Fig. 6. This list of "related stories"
is generated by taking the closed-captions of the currently opened story as
query text and the top ten results arereturned at browsing time. Thus, in
Fig. 6 the related stories shown may be again stories about Paul Bremer but
may be about something else. In this way, the user can start by browsing
a news story detail, followed by jumping between related stories that the
system automatically generates links to. Users can also access the news stories
by automatic recommendation in which they indicate their preference for a
particular news story using a 5-point thumbs-up and -down scale icons located
beside each story, and as this information from the users accumulates over time

Fig. 6. Browsing Related Stories

the system can recommend some of the newly appeared stories as well as older stories in the archive to individual users by way of collaborative filtering [245].

A more elaborate presentation and interaction scheme to support an effective news story navigation would be possible on top of the interface that current Físchlár-News features, for example a timeline presentation of related stories [317], visualising the topic thread over time [153], use of clustering techniques to visualise clusters of topically related stories and to highlight recently added stories in each of the clusters [74], use of time and locality to automatically generate an interactive collage of images, video and text of news stories presented with timeline and map [61], and visualisation of news themes in a virtual thematic space where a user can navigate through appearing and disappearing textual themes in a highly interactive way [262]. These potentially useful presentation techniques will require more investigation on their usability before deployed and used by real users.

Físchlár-News has been operational on our University campus for over 3 years continuously capturing, processing and archiving daily TV news, to support media and journalism students and staff as well as other users who want casual news updates during the day. Apart from serving as an experimental platform for the various content-based analyes described above, the value of the system over watching the news on TV is that using these techniques the system automatically turns the sequential, time-based medium of many hours of news video (currently several hundred hours of video content) into an easily browseable and searchable commodity which allows convenient story-based access at any time. More details of technical aspects of the system and its envisaged usage scenario have been drawn in [298], and a long-term user study on people's actual usage of the system in the workplace can be found in [195]. There are many commercial online news websites that are highly up-to-date and feature photos and video footage with links to related stories, but their human indexing and authoring means a high cost of manual work and a difficulty in maintaining indexing consistency among indexers over time.

4 Físchlár-TRECVid: System Description

The Físchlár-News system, that we have just described, is an operational video retrieval system with a campus-wide user base, hence the user interaction supported is clearly defined and easy for any user to understand and operate. Físchlár-TRECVid, on the other hand, is an experimental retrieval system, the aim of which is to evaluate alternative techniques to interactive video search and retrieval. This evaluation is conducted annually as part of the TRECVid workshop.

4.1 TRECVid: Benchmarking Video Information Retrieval

The history of (text-based) information retrieval is one where empirical investigation and experimentation has always been fundamental. Information retrieval draws its background from a combination of computer science, information science, engineering, mathematics, human-computer interaction and library science and throughout its 40 years of history theoretical improvements have always had to be validated in experiments before being accepted to the IR community. This philosophy has led to the emergence of the annual TREC (Text REtrieval Conference) exercise which, since 1991, has facilitated the comparative evaluation of IR tasks in an open, metrics-based forum. TREC is truly global with 100+ participating groups in 2004.

In 2001, TREC featured yet another "track" or activity, on tasks related to video information retrieval, including shot boundary detection, feature extraction, and interactive searching. This has now spun out as a separate independent exercise known as TRECVid [325, 300]. The operation of TRECVid, and TREC, revolves around the organisers, National Institute of Standards and Technology (NIST) gathering and distributing video data to signed-up participants (60+ participants in 2005). In 2001 this was only 11 hours of video content and in 2005 it is 200+ hours of video. Although the tasks of shot and story bound detection and of feature identification are of importance to video IR, we are interested in the interactive search task here, where users are given a multi-modal (multiple media) topic as an expression of an information need and a fixed time window (15 minutes usually) to complete the task of finding as many shots likely to be relevant to the topic as they can. Note that the user task is to locate shots, not news stories, which are likely to be relevant.

TRECVid participants use the same video data, run the same topics (descriptions of an information need) against this using their own systems and then the relative performance of systems/groups in terms of retrieval effectiveness is measured. The TRECVid excercise has participation from dozens of research groups worldwide and is a true benchmark of the effectiveness of different approaches to video retrieval. When NIST receive the identified shots from each participating site and for each topic, these are then pooled together and duplicates eliminated. Remaining shots are then presented to assessors who examine each retrieved shot and then make a binary judgement as to relevance. This establishes the ground truth for each of the topics and with this information available, the organisers are then able to measure the performances of the retrieval runs submitted by each participating site and compute retrieval performance figures in terms of precision and recall, for each.

In the 2004 edition of TRECVid, the video data distributed to participating groups was broadcast TV news, from CNN and ABC. The interactive search task was to retrieve *shots* which matched the topic, not news stories and the nature of the search topics illustrates this. Topic 144 asks users to "Find shots of Bill Clinton speaking with at least part of a US flag visible behind him" and Fig. 7 shows the 2 images and keyframes from the 2 video clips which form part of the topic definition. Associated with the TRECVid 2004 broadcast data there were three types of text information distributed to participants; the original closed-captions which give an accurate summary but not an exact replication of the dialogue as spoken in the video, the output from an automatic speech recognition system, and "video OCR" which corresponds to the character recognition of any text appearing in the video frame, such as a sub-title or text overlay.

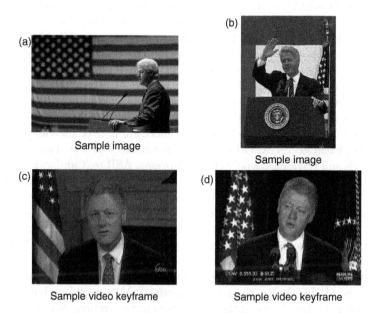

Fig. 7. Sample images ((a) and ((b)) and keyframes from sample video clips ((c) and (d)) for topic 144 from TRECVid 2004.

Finally, several groups evaluated their own feature (semantic concept) detection algorithm(s) for each of the 33,367 shots in the collection and submitted the results of their analysis to NIST as part of one of the TRECVid tasks, but also made the results of these feature detections available to other participants for use in their own search systems (distributed in MPEG-7 format). The features whose detection performance was evaluated and whose results were used in some TRECVid search systems, including our own, were boats or ships of any type, the presence of Madeleine Albright, the presence of Bill Clinton, trains or railroad cars, a beach with the water and the shore visible, a basketball score with the ball passing down the hoop and into the net, an airplane taking off, people walking or running, physical violence between people and/or objects, and finally, a road of any size, paved or not. Some of these are really difficult to do and represent very challenging tasks while others are more achievable. Some of the donated features were used by us in the system described below and by other participating groups in their interactive retrieval experimentation.

In only 5 years of operation TRECVid has grown considerably in terms of the data volume, the number of groups taking part, the tasks being evaluated, the measures used and the complexity of the whole exercise. It is within this framework that we developed a version of our Físchlár system for TRECVid in 2004, which we call Físchlár-TRECVid, which we now describe.

4.2 Físchlár-TRECVid

In Físchlár-TRECVid, search and retrieval of video shots is supported as well as the browsing of news programmes at the shot level. The shot retrieval engine employed for Físchlár-TRECVid is based on a combination of query text matched against spoken dialogue combined with image-image matching where a still image (sourced externally), or a keyframe (from within the video archive itself), is matched against keyframes from the video archive. The image matching is based on low-level features taken from the MPEG-7 eXperimentation Model (XM) [354].

Unlike Físchlár-News, which operates successfully over the closed-caption text alone, Físchlár-TRECVid employs three sources of text data to support shot search and retrieval namely closed-captions, ASR text and video OCR. It has been shown [68] that the integration of multiple sources of text improves retrieval performance over using a single source of text transcripts alone, when operating over TV news programmes. For example, the addition of closed-caption and OCR text to an existing automatic speech recognition transcript-only retrieval engine improves searching performance by 17% (MAP) and the number of relevant shots found for a typical query by 18%.

As stated earlier, the visual shot matching facilities were primarily based on using the MPEG-7 eXperimentation Model (XM) to provide shot matching services over the keyframes of the video shots in the archive. We incorporated four XM algorithms: local colour descriptor; global colour descriptor; edge histogram descriptor and homogenous texture descriptor. This allowed us, for a given shot (using the representative keyframe) or externally sourced query image, to generate a ranked list of shots that best match a query image. The user of the system was allowed to choose which of the XM techniques were to be used for any given query.

In addition to these four visual shot matching techniques, we also incorporated two additional image processing techniques to improve visual shot matching performance. The first of these was motion estimation which allowed us to rank shots from the collection with regard to similarity of motion within the shots and the second was a face filter which filtered out shots from the video archive that contain one or more faces. This would be very useful, for example, if a query was to find video of people or known persons. Of course, in a video search and retrieval system such as Físchlár-TRECVid which supports both text and image based retrieval, the facility to query using both text and image evidence is essential, especially if relevance feedback is to be supported. In our experience (and for other TRECVid participants as well), the addition of visual shot matching techniques to a text-based video retrieval system improves retrieval performance, in our experiments by 13% [68].

Similar to Físchlár-News, which supports a version of relevance feedback in the form of its "related stories" and recommendation features, Físchlár-TRECVid supports relevance feedback at the shot level. For example, if a user queries for "forest fires" using a text-only query and locates a number of good examples of shots of forest fires, then one or more of these can be added to the query and included in subsequent searches. In this way a query can be augmented and refined by adding relevant shots to the query as the user locates and identifies these shots. As shots are added to the query, the most important terms from the shots are extracted and used to augment the text aspect of the user query and the keyframes from these shots are employed for visual shot matching.

The interface to Físchlár-TRECVid is shown in Fig. 8 and is comprised of three panels. On the left of the screen is the query panel and below this is the playback window, in which the video from any selected shot can be played back through the video player. On the right of the screen is the saved shots panel in which the user keeps track of shots that have been found to be relevant. In the centre is the search result panel which displays the results of user interaction and search. As can be seen (in Fig. 8) a query which is comprised of the text 'rocket launch' and a single sample image has been entered. In response to this query, the user has been presented with the ranked list of groups of shots (the top three shown), twenty per page with a total of five pages of results available.

Presenting results from an archive of news stories is ideally done at the news story level. So a user looking for news concerning, for example, President Bush, would be presented with a ranked list of news stories about George Bush, as is the case with Físchlár-News. However, when operating at the shot level, there may be many useful video shots adjacent or near to a highly ranked shot, but which themselves may not be ranked highly. This is because the spoken text or closed-caption text may not be precisely synchronised with the

Fig. 8. Searching using text and image examples

video on the screen at that point. Our experience of running user experiments into interactive video search and retrieval suggests that relevant shots are often found adjacent to the highest ranked shots, especially when the query is composed of text alone. To overcome this problem results are presented not as shots but rather as groups of adjacent shots (see the centre panel in Fig. 8) in which three results are displayed, each of which is composed of five adjacent shots, this giving the user the context of each result. Within this group of five adjacent shots, the middle shot is the matched shot for the group, with the adjacent shots providing context and only slightly influencing the shot ranking of the matched shot. The matched shot's keyframe is displayed largest (with a red border) and the neighbouring two keyframes are progressively smaller. In addition, the speech recognition transcript text of the five shots is presented immediately below the shots with the query terms highlighted to provide additional context. The two buttons below each keyframe in the 'search result' panel allow the user to either add the shot to the query or add the shots to the collection of saved shots for a given query.

Nearby a matching shot or somewhere within the broadcast, there is a likelihood that there will be more relevant shots. By providing a mechanism to see the full broadcast for any given shot, the system can allow more efficient searching. However, going into a full broadcast (25-30 minutes long) is at the same time making the user's browsing space considerably larger, and thus will require more user effort. Our experiences suggest that the user is less likely to browse a full programme than scan a ranked list of results. Taking this onboard, in Físchlár-TRECVid, we support a three-level search and browse hierarchy. In addition to the first level results (Fig. 8) from all programmes, the user can see all matched shots within a broadcast (by selecting the "MORE MATCHES IN THIS BROADCAST" option immediately above the five grouped keyframes). This will present a ranked list of groups of shots from a given broadcast and thus aids the user in locating further useful shots without having to browse all the keyframes in the programme. That said, if the user thinks there could be further more matches within other parts of this broadcast, she clicks on "BROWSE FULL BROADCAST" link from the second level (not shown), which will bring the user into the full broadcast browsing which is the third level of our three-level hierarchy. In full broadcast browsing, an interactive SVG (Scalable Vector Graphics) timeline is presented with the matched points in the broadcast highlighted, to allow the user to quickly jump within the broadcast (shown at the top of the result panel in Fig. 9). Our experiences from user experiments suggest that the three-level hierarchy is beneficial in that a user only need browse all keyframes from a full programme when the previous two levels suggest that there may be more relevant content to be found at the programme level.

As we have mentioned, at any point in the search session, shots (represented by keyframes and associated text) can be added to the query in a process of relevance feedback. In Fig. 9 there are two shots added to the

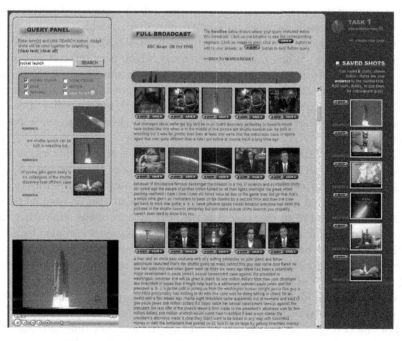

Fig. 9. Browsing a full broadcast (with saved shots)

query (in the query panel on the left of the screen) in addition to the original query image. The text associated with these two shots is displayed beside their keyframes. The 'search result' panel now shows a user browsing an entire broadcast, which presents a temporally organised listing of keyframes and the text transcript from these shots.

After a phase of search and browse, the 'saved shots' panel (rightmost panel of Fig. 9) will contain shots that the user considers relevant to the information need. These shots can be added to the query for further relevance feedback, or removed from the 'saved shots' panel. More details of the technical apsects of Físchlár-TRECVid, a more detailed user iteraction scenario and its comparative performance are presented in [68].

5 Analysis of Video Searching and Browsing

In the case of Físchlár-News and the Físchlár system developed for our participation in TRECVid in 2004, we have seen how they analyse and index video and support different ways for a searcher to navigate video archives. Físchlár-News supports searching and browsing video where the unit of information is the TV news story. Each news broadcast can contain between 10 and 20 individual stories and Físchlár-News is designed and built to support people searching for news information. Sometimes users want to get a high level gist of *all* the news on a given day in which case browsing the archive by calendar and seeing a summary of all news stories on a given date is sufficient. Other times a user wants to search for news on a particular topic, in which case a keyword search against the spoken dialogue will result in a ranked list of stories for the user to browse. When a user finds a story which is of interest and wants to locate other stories on the same topic then the automatically-generated links to related stories provides this. As we have shown in an extensive user study of Físchlár-News [195], the system has enough functionality to satisfy its users as a tool for searching and browsing TV news video on a regular basis.

The Físchlár-TRECVid system supports users searching for video *shots* using a combination of text from the closed-captions or the automatic speech recognition, and/or using sample images which in some way illustrate or capture the information need, collectively. Video clips can also be used as part of the search criteria but in this case it is the keyframe from the clip rather than the clip per se, that is used. Relevant or useful video shots can also be added to the search and used as part of an expanded search which combines text searching and video/image searching into one. In searching for video shots in Físchlár-TRECVid, raw metadata such as date, location or program name does not offer any kind of useful support for searching and is not used since the information need addressed is entirely *content-based*.

For the two systems the user needs addressed are very different, and thus the two sets of specific functionalities offered are different. Browsing among news stories in Físchlár-News is very rapid and users can easily jump from one story to another in several ways. Browsing among shots in Físchlár-TRECVid is also equally fast with support for rapid visualisation across shots and the rapid location of required shots.

Searching in both systems uses text derived from closed-captions or speech recognition, and this is sufficient where the search is for information which appears in the dialogue as in TV news. Where the information need is partly based on what appears in the video then we use some aspects of image searching by extracting visual features from the video content. However, the visual features we use are low-level characteristics like colour and texture , but these have no real semantic value and higher-level feature detectors, such as airplane taking off, or beach scenes, have proved difficult to detect with a high degree of accuracy. To move beyond feature extraction from an *entire* video frame as a basis for image searching we need to use sub-parts of an image, and ideally these should be the major objects appearing within a frame, or within a shot. There has been recent preliminary work on using video objects as a basis for video retrieval in [165, 282] and while this shows promise and appears to work well it has yet to be tried on really large collection sizes. Object-based video retrieval is one of the key challenges and areas for future research, but the main hurdle to achieving this remains the automatic identification of video objects, which has been a challenge to the video analysis community for some time and can currently be done only semi-automatically [2].

With more than one search option available (closed-captions, ASR, various low-level visual features like colour and texture, and higher-level features and possibly even video objects) one issue is how should we use these different search options in combination. A recent study of different combination methods has provided some insights [178] but the best approach appears to be to blend different search types together in a weighted combination where the weights depend on the type of query [139] as used by the Informedia system at TRECVid 2004.

However, in general we can say that while we use image features in video retrieval, we don't really use much of the visual features of video in video retrieval. We use keyframes only and we rarely use the temporal aspect of video, no inclusion of camera motion, no inclusion of object motion (though there are exceptions [261]) and so we have a long way to go in video search to develop it to a comparable level as, say, web searching.

6 Conclusions

It is inevitable that content-based information retrieval, including searching, browsing, summarisation and highlight detection of video information, is set to become hugely important as video becomes more and more commonplace. During 2003 alone, Google ran 50 billion web page searches and during 2004 AskJeeves ran more than 20 million web page searches per day, globally.[2]

---

[2] Tuic V. Luong, Sr. VP Engineering and Technology, AskJeeves, at the 9th SearchEngine Meeting, The Hague, The Netherlands, 19-20 April 2004.

These figures indicate how embedded the web and web *searching* have become into our society. If video is to become even a fraction as important as we believe it is, then video searching and video browsing are critical technologies.

At the present time the development of effective video IR is decades behind text-based IR, but is catching up fast. What will accelerate this is what has accelerated web page searching, namely commercial interest, and it will be across a range of video genres and a range of applications. Searching CCTV for possible suspects, searching broadcast TV news archives for past stories about tropical storm damage in Florida, searching a person's recorded TV programs on their own DVR, searching an online archive of past episodes of "Here's Lucy" to find the hilarious scene where she is working in a jam factory or searching through personal (home) video to find clips of your son and daughter together over different family vacations. These are all examples of the kind of searches we will want to do, and the need for which will drive the development of video IR.

From our experiences we can conclude that video navigation consists of search (with relevance feedback being of high importance), local browsing and collection-wide link traversal. The search techniques employed rely heavily on old text search technology with some help from visual shot matching techniques. However the video search technology could be so much more. Over the coming years we will see many advances in video search and retrieval in a number of 'hot topics'. For example, the visual shot matching techniques we have employed in the research presented in this chapter are still at the early stages of development. A user does not intuitively think of an image or video in terms of colours, textures and edges or shapes, rather the user understands semantic concepts (cars, explosions, etc.) and would like to query using these. Hence, object based search and retrieval will be a key technology where a user can define and select an object from a video clip and search for that object across an entire archive. Also key advances will come in the area of security and intelligence, where huge archives of digital video footage will be gathered, indexed and objects identified, which in so doing will help to solve another video IR problem at present; that of searching extremely large archives of tens or hundreds (or more) of thousands of hours of video content. What the search engine has done for text retrieval, security and intelligence requirements may do for digital video retrieval. Other 'hot topics' include summarisation and personalisation of video content, so that a user only gets a summary of important or novel video, in response to a query or information need.

Acknowledgements

This work is supported by Science Foundation Ireland under grant 03/IN.3/I361. The support of the Enterprise Ireland Informatics Directorate is also gratefully acknowledged.

Locating Information in Video by Browsing and Searching

Andreas Girgensohn[1], Frank Shipman[2], John Adcock[1], Matthew Cooper[1], and Lynn Wilcox[1]

[1] FX Palo Alto Laboratory, 3400 Hillview Ave. Bldg. 4, Palo Alto, CA 94304, USA.
{andreasg,adcock,cooper,wilcox}@fxpal.com
[2] Department of Computer Science & Center for the Study of Digital Libraries
Texas A&M University, College Station, TX 77843-3112, USA.
shipman@cs.tamu.edu

1 Introduction

Locating desired video footage, as with other types of content, can be performed by browsing and searching. However, unlike browsing textual or image content, this task incurs additional time because users must watch a video segment to judge whether or not it is relevant for their task. This chapter describes browsing and searching interfaces for video that are designed to make it easier to navigate to or search for video content of interest.

Video is typically viewed linearly, and navigation tools are limited to fast forward and rewind. As a result, many approaches to summarize videos have been proposed. One approach is to support skimming via playback of short versions of the videos [63, 206, 315]. Another approach is to support access to video segments via keyframes [358] Finally, video libraries let users query for video segments with particular metadata, e.g., topic, date, length [221] We are exploring two alternative approaches. The first approach utilizes interactive video to allow viewers to watch a short summary of the video and select additional detail as desired. Our second approach lets users search through video with text and image-based queries. This type of video search is difficult, because users need visual information such as keyframes or even video playback to judge the relevance of a video clip and text search alone is not sufficient to precisely locate a desired clip within a video program.

1.1 Detail-on-demand Video

We use detail-on-demand video as a representation for an interactive multilevel video summary. Our notion of detail-on-demand video has been influenced by interactive video allowing viewers to make choices during playback impacting the video they subsequently see. An example of interactive video is DVDs that include optional sidetrips that the viewer can choose to take. For example, when playing "The Matrix" DVD with optional sidetrips turned on, the viewer sees a white rabbit icon in the upper left corner of the display that indicates when a link may be taken. These links take the viewer to video segments showing how the scene containing the link was filmed. After the sidetrip finishes playing, the original video continues from where the viewer left off.

Our interactive multi-level video summary takes the form of a hypervideo comprising a set of video summaries of significantly different lengths and navigational links between these summary levels and the original video. Viewers can interactively select the amount of detail they see, access more detailed summaries of parts of the source video in which they are interested, and navigate throughout the entire source video using the summary. This approach explores the potential of browsing via hyperlinks to aid in the location of video content.

1.2 Video Search

For situations where browsing is not practical due to the size or semantic structure of the video collection, we describe a system for searching with text and image-based queries. While searching text documents is a well-studied process, it is less clear how to most effectively perform search in video collections. In text search it is typical for documents to be the unit of retrieval; a search returns a number of relevant documents. The user can then easily skim the documents to find portions of interest. In cases where documents are long, there are techniques to identify and retrieve only the relevant sections. [327].

However, treating entire video documents or programs as units of retrieval will often lead to unsatisfactory results. After retrieving relevant programs, relevant clips, typically less than one minute of length, must be located within their source videos. Even when such videos are broken into sections, or stories of several minutes in length, it is still time-consuming to preview all the video sections to identify the relevant clips.

Our approach to this problem is to provide flexible user interfaces that present query results in a manner suited for efficient interactive assessment. Our target users are analysts who need both visual and textual information or video producers who want to locate video segments for reuse. While the latter will frequently use libraries that support retrieval with extensive meta-data describing properties such as location, included actors, time of day, lighting conditions, our goal is to enable search within video collections where such meta-data is not available. In this work, we assume that time-aligned text, such as transcripts, automatically recognized speech, or closed captions, is available.

The next section provides a conceptual overview of our approaches for detail-on-demand video and video search and compares these approaches to related work.

2 Overview

We support two approaches for the navigation of video collections, detail-on-demand video and and interactive video search. Both approaches have similarities to other approaches described in the literature but emphasize rich interfaces and intuitive interaction over complexity in analysis and interaction.

2.1 Detail-on-demand Video

Hypervideo allows viewers to navigate within and between videos. Applications of hypervideo include educational environments [121] and narrative storytelling [283]. General hypervideo allows multiple simultaneous link anchors on the screen, e.g., links from characters on the screen to their biographies. This generally requires anchor tracking — tracking the movement of objects in the video that act as hotspots (source anchors for navigational links) [143, 303] We have chosen to investigate detail-on-demand video as a simpler form of hypervideo, where at most one link is available at any given time. Such video can be authored in a direct manipulation video editor rather than requiring scripting languages or other tools that are unsuitable for a broad user base. At its simplest, the author selects a segment of the edited video for which a link will be active and the video sequence that will be shown if the viewer follows that link. By removing the need to define and track hot spots in video, the authoring and viewing interfaces can be simplified.

The name detail-on-demand comes from the natural affordances of this form of hypervideo to support gradual access to specific content at finer granularities. With the main video stream presenting the topic or process at a more abstract or coarser-grained level, the viewer can navigate to view clips on the topic of interest. This improvement can save the viewer's time in comparison to current linear access to videos or video summaries. The representation's primary features are navigational links between hierarchical video compositions and link properties defining link labels and return behaviors.

Detail-on-demand video consists of one or more linear video sequences with links between elements of these sequences (see Fig. 1). Each video sequence is represented as a hierarchy of video elements. Segments of source video (clips) are grouped together into video composites, which themselves may be part of higher-level video composites. Links may exist between any two elements within these video sequences. The source element defines the source anchor for the link — the period of video playback during which the link is available to the viewer. The destination element defines the video sequence that will be played if the viewer follows the link. The source and destination elements

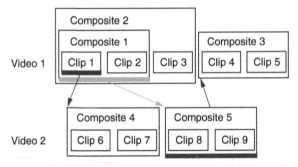

Fig. 1. Hierarchically-organized videos and links

specify both a start and an end time. To keep the video player interface simple, detail-on-demand video allows only one link to be attached to each element. For more information on this representation see [292].

A variety of interfaces for accessing video make use of an explicit or inferred hierarchy for selecting a starting point from which to play the video. These vary from the standard scene selection on DVDs to selection from hierarchically structured keyframes or text outlines in a separate window [274, 230]. Selecting a label or keyframe in a tree view is used to identify a point for playback.

The primary difference between interfaces supporting hierarchical access to video and detail-on-demand video is that the detail-on-demand viewer may request additional detail while watching the video rather than having to use a separate interface such as keyframes or a tree view. Also, the hierarchical representation of a video may not include semantics beyond simple hierarchical composition. The links in hypervideo have labels and a variety of behaviors for when the link's destination anchor finishes playback or when the user interrupts playback. There are two independent link behaviors to define: (1) what happens when the destination sequence of a video link finishes playing, and (2) what happens when the viewer of the destination sequence ends its presentation before it is finished. The four options are (1) play from where the viewer left the video, (2) play from the end of source anchor sequence, (3) play from beginning of source anchor sequence, and (4) stop playback.

Links in hypervideo serve many of the same purposes they serve in hypertext. They provide opportunities for accessing details, prerequisite knowledge, and alternate views for the current topic. Unlike hypertext, the effectiveness of the hypervideo link is affected by link return behaviors, or what happens after the destination video has been watched or its playback aborted.

2.2 Video Search

Video search is currently of great interest, as evidenced by recently unveiled web-based video search portals by Yahoo [353] and Google [117]. 2004 marked the fourth year of the TRECVID [324] evaluations which draws a wide variety of participants from academia and industry. Some of the more successful ongoing efforts in the interactive search task draw upon expertise in video feature identification and content-based retrieval. The Dublin City University effort [69] includes an image-plus-text search facility and a relevance feedback facility for query refinement. The searcher decides which aspects of video or image similarity to incorporate for each query example. Likewise, the Imperial College interactive search system [141] gives the searcher control over the weighting of various image features for example-based search, provides a relevance feedback system for query refinement, and notably incorporates the NN^k visualization system for browsing for shots "close" to a selected shot. The MediaMill, University of Amsterdam system [307] is founded on a powerful semantic concept detection system allowing searchers to search by concept as well as keyword and example. Likewise the Informedia system from Carnegie Mellon University [62] incorporates their mature technology for image and video feature detection and puts the searcher in control of the relative weighting of these aspects.

Our effort is distinguished from others primarily by the simplicity of our search and relevance feedback controls in favor of an emphasis on rich interfaces and intuitive paths for exploration from search results. Our scenario is not so much one of query followed by refinement as it is query followed by exploration.

Whether explicitly stated or not, a goal in all of these systems is a positive user experience. That is, an informative and highly responsive interface cannot be taken for granted when handling thousands of keyframe images and tens of gigabytes of digital video.

The next two sections present two systems for browsing and searching video data. Section 3 presents a system that summarizes longer videos at multiple granularities. The summaries are combined to form detail-on-demand video enabling users to explore video and locate specific portions of interest. Section 4 presents our system for searching a video repository using text, image, and video queries. The repository's content is automatically organized at the program, story (i.e. topic), and shot levels enabling a powerful interface. We compare both approaches to related work. We summarize the paper in Sect. 5.

3 Hypervideo Summarizations

For many types of video, simply viewing the content from beginning to end is not desired. For example, when viewing an instructional video, you may want to skip much of the material and view only a topic of particular interest. Similarly, when viewing home video you may want to watch only particular sections. Detail-on-demand videos can be structured as an interactive summary providing access into longer linear videos.

Fig. 2. An automatically generated hypervideo summary in Hyper-Hitchcock

We created the Hyper-Hitchcock system as a means to author detail-on-demand videos. The Hyper-Hitchcock authoring interface (see Fig. 2) allows the rapid construction of hierarchies of video clips via direct manipulation in a two-dimensional workspace. Navigational links can be created between any two elements in the workspace. A more detailed description of the Hyper-Hitchcock authoring interface can be found in [292].

Human authoring of such summaries is very time consuming and not cost effective if the summary will only be used a few times. We developed an algorithm for the automatic generation of hypervideo summaries composed of short clips from the original video. The interactive video summaries include linear summaries of different lengths and links in between these summaries and the entire source video.

The following sections describe the user interface for viewing hypervideo summaries, different clip selection techniques that are suitable for different types of video (e.g., home video or produced training video), and methods for placing links and setting link behaviors that improve interaction with the hypervideo summary.

The generation of the multi-level video summary includes three basic decisions: (1) how many levels to generate and of what lengths, (2) which clips from the source video to show in each level of the summary, and (3) what links to generate between the levels of the summary and the behaviors of these links.

3.1 Viewing Interface

We have created a series of viewing interfaces for Hyper-Hitchcock [114]. These interfaces vary in their complexity and the requirements they place on the player. The most recent, shown in Fig. 3, is the result of a series of user studies [291].

We found that visualizing links in the timeline without labels or keyframes was confusing to users. However, putting link labels inside the timeline caused confusion between the actions of link following and skipping within the current video sequence. To address these issues, we placed keyframes for all links along the timeline and make all links available via keyframes and not just the link attached to the currently playing video segment. Users can follow the links by clicking on link keyframes or labels without having to wait for or move the video playback to the appropriate place. The keyframes for inactive links are reduced in size with faded link labels. The area of the active link, defined to be the innermost source anchor currently playing, is emphasized in the timeline and separators between links in the timeline are made stronger. The keyframe for the link under the mouse is also enlarged and the link is emphasized even more than the active link to indicate that mouse clicks will select it.

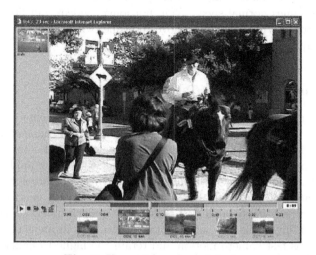

Fig. 3. Hypervideo player interface

We display a stack of keyframes representing traversal history in the top-left of the player window. The keyframes for the older jump-off points are scaled down to enhance the history view. All keyframes are clickable, thus enabling the user to backtrack through several links at once.

3.2 Determining the Number of Summary Levels

The number and length of summary levels impact the number of links the user will have to traverse to get from the top-level summary to the complete source video. Having more levels provides users more control over the degree of summarization they are watching but also makes access to the original video less direct. To reduce the user's effort in navigation, a system can allow the user to traverse more than one link at once (moving more than one level deeper into the hypervideo).

Our current approach to determining the number of levels in the interactive summary is dependent on the length of the source video. For videos under five minutes in length, only one 30-second summary level is generated. For videos between 5 minutes and 30 minutes, two summary levels are generated — the first level being 30 seconds in length and the second being 3 minutes in length. For videos over 20 minutes, three summary levels are generated — one 30 seconds long, one three minutes long, and the last being one fifth the length of the total video to a maximum of 15 minutes. The length of the lowest summarization level is from one fifth to one tenth the length of the original video, except in cases of very short (less than two and a half minutes) or very long (more than 150 minutes) original videos.

This summarization ratio provides value over viewing a time-compressed presentation of the source video. Wildemuth and colleagues found significant performance drop-offs in comprehending a video when fast forward rates rose above 32–64 times normal speed [339]. For longer videos, our 30-second top-level summary provides speedups above 100-times with the option to see more detail for any interesting portion.

3.3 Segmenting Video into Takes and Clips

Our algorithms assume that the video has been first segmented into "takes" and "clips". For unproduced (home) video, takes are defined by camera on/offs. We do most of our work with DV format video that stores the camera on/off information, but when this metadata is absent we can apply automatic shot boundary determination (SBD) methods [70]. Clips are subsegments of takes generated by analyzing the video and determining good quality segments [113]. Here, good quality is defined as smooth or no camera motion and good lighting levels. We segment takes into clips in areas of undesirable quality such as fast camera motion and retain the highest-quality portion of the resulting clips.

For produced video, takes are defined as scenes and clips are the shots of the video. These are identified using well-known techniques [70, 367]. Sundaram and Chang [315] propose using consistency with regard to chromaticity, lighting, and ambient sound as a means for dividing a video into scenes. We are still experimenting with segmenting produced video but we plan to use a variation of published approaches.

3.4 Selecting Clips for Summary Levels

We have explored a number of algorithms for selecting clips from original source video. Selection of clips to use for each video summary is closely related to traditional video summarization. Unlike traditional summarization, selection of clips not only effects the quality of the generated linear summary but also impacts the potential for user disorientation upon traversing and returning from links.

We will describe three clip selection algorithms we have developed in this section. The first two approaches, the clip distribution algorithm and the take distribution algorithm, described below select clips based on their distribution in the video. They are geared for unproduced, or home video, where clips have been selected by their video quality. The third approach, the best-first algorithm, assumes that an external importance measure has been computed for the clips (shots) [63, 327]. This algorithm is more suitable for produced video.

The clip distribution algorithm is based on the identification of an array of m high-quality video clips via an analysis of camera motion and lighting. The current instantiation assumes an average length of each clip (currently 3.5 seconds) so the number of clips n needed for a summary is the length of the summary in seconds divided by 3.5. The first and last clips are guaranteed to be in each summary with the remainder of the clips being evenly distributed in the array of potential clips. The use of an estimate of average clip length generates summaries of approximately the desired length rather than exactly the requested length. The algorithm can easily be altered to support applications requiring summaries of exact lengths by modifying in/out points in the selected clips rather than accepting the in/out points determined by the clip identification algorithm.

Applying the clip distribution algorithm will cause the resulting summary levels to include more content from takes that include more clips. This is appropriate when the camera is left filming while panning from activity to activity, such as walking from table to table at a picnic or reception. If shot divisions are not reflective of changes in activity then a take is likely to be over or underrepresented in the resulting summary. For example, a single take that pans back and forth between two areas of a single activity (people having a discussion or sides of the net in a tennis match) is likely to be overrepresented. Similarly, a single take consisting of several activities filmed back-to-back at a particular location (e.g., video of a classroom lecture) is likely to be underrepresented.

The take distribution algorithm uses the same segmentation of the video of length L into takes and clips but balances the representation of takes in the summary. For the first level, set a length L_1 (e.g., 30 seconds) and a clip length C_1 (e.g., 3 seconds) to pick $n = (L_1/C_1)$ clips. Check the centers of intervals of length L/n and include a clip from each of the takes at those positions. Pick the clip closest to the interval center. If more than one interval center hits the same take, pick the clip closest to the center of the take. If fewer than n clips are picked, look for takes that have not been used (because they were too short to be hit). Pick one clip from each of those takes starting with the clip that is furthest away from the already picked clips until n clips are picked or there are no more takes that have not been used. If still fewer than n clips are picked, pick an additional clip from each take in descending order of the number of clips in a take (or in descending order of take duration) until enough clips are picked. Continue picking three and more clips per take if picking two clips per take is insufficient. The same method can be used for the second level with lengths L_2 (e.g., 180 seconds) and clip length C_2 (e.g., 5 seconds).

The take distribution algorithm emphasizes the representation of each take in the summary and the distribution of selected clips throughout the duration of the source video. This approach will provide greater representation of short takes as compared to the clip distribution algorithm. This is advantageous when a consistent number of distinct activities are included in takes. It is likely to underrepresent takes in cases where many activities are included in some takes and one (or few) are included in others. Different application requirements will make one or the other algorithm more suitable. The take distribution algorithm will better represent the tennis match or conversation but the clip distribution algorithm better reflects the case of a camera moving between picnic tables.

Both of the above algorithms are designed to provide glimpses into the source video at somewhat regular intervals — with the first algorithm using the number of clips (or shots) as a measure of distance and the second algorithm using takes and playing time as a measure of distance. In contrast, the best-first algorithm for selecting clips uses an importance score for clips to select the most important video first. Importance scores for clips can be assigned automatically, using heuristics such as clip duration and frequency, as in [327]. Alternatively, scores can be assigned manually using the Hyper-Hitchcock interface. For example, in an instructional video, longer clips showing an entire process would be given a higher importance than shorter clips showing a particular subprocess in more detail. To generate a best-first summary, clips are added to the summary in order of their importance. This results in each level being a superset of higher levels (shorter) of the summary.

3.5 Placing Links Between Summary Levels

Once a multi-level summary has been generated, the next question is which links to generate between the levels. Links take the viewer from a clip at one level to the corresponding location in the next level below. In general, links are created in the multi-level summary for viewers to navigate from clips of interest to more content from the same period.

Generating links includes a number of decisions. A detail on demand video link is a combination of a source anchor, a destination anchor and offset, a label, and return behaviors for both completed and aborted playback of a destination. We will describe two variants for link generation — the simple take-to-take algorithm, and the take-to-take-with-offsets algorithm — to discuss trade-offs in the design of such techniques.

The simple take-to-take link generation algorithm creates at most one link per take in each level of the hypervideo. Figure 4 shows an example of a three-level summary created from 23 high-value clips identified in a four-take source video using this approach. All the clips from a particular take are grouped into a composite that will be a source anchor for a link to the next level. A composite in a higher-level summary (shorter) will be linked to the sequence of clips from the same take in the next level. If a take is not represented in a higher level, it will be included in the destination anchor for the link from the previous take. Otherwise, there would be no way to navigate to the clips from that take in the lower summary level. For example, let us consider the link from the middle clip in the top level of the summary shown in Fig. 4. In this case, Clip 12 in Level 1 is the source anchor of the link. The destination anchor is a composite composed of Clip 10 and Clip 15. Clip 15, which is from Take 3, has been included because there was no clip from Take 3 in Level 1.

The take-to-take-with-offsets algorithm depicted in Fig. 5 is similar to the simple take-to-take algorithm except that a separate link is generated for each clip in the source level of the summary. Clips are grouped into composites as in the above algorithm but links for clips other than the first clip representing a take in a particular level will include an offset to take the viewer to the (approximately) same point in the next level rather than returning to the first clip for that take in the next level.

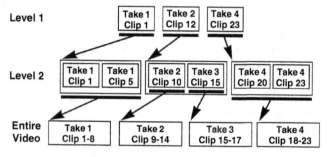

Fig. 4. Links generated by the simple take-to-take algorithm for a three-level video summary

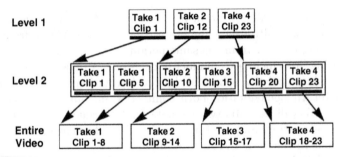

Fig. 5. Links generated by take-to-take-with-offsets algorithm for a three-level video summary

The addition of more links is meant to keep users from having to re-watch video and to provide more precise temporal access into the video. The described method for setting offsets minimizes the "rewinding" that automatically occurs when a link is taken. If users do not realize they want material from this take until they view later clips, then they have to "rewind" the video by hand. An alternative method to setting offsets that is in between the no-offset approach of the simple take-to-take link generation and setting the offset to the same clip is to view the clip sequence as a timeline and take the user to the first clip that is closest to the clip being played. This would take the user to the midway point between the last available link and the start of the Clip being played when the link is selected.

The simple take-to-take link algorithm works well when the summary levels are being used as a table of contents to locate takes within the source video. In this case, the user decides whether to get more information about a take while watching a montage of clips from that take. A difficulty with the simple take-to-take algorithm is that links to takes that include many activities will take the user to the take as a whole and not any particular activity. The take-to-take-with-offsets algorithm allows more precise navigation into such takes but assumes the user can make a decision about the value of navigation while viewing an individual clip.

Currently, link labels in the hypervideo summary provide information about the number of clips and length of the destination anchor. Algorithms that generate textual descriptions for video based on metadata (including transcripts) could be used to produce labels with more semantic meaning.

Link return behaviors also must be determined by the link generation algorithm. When the user "aborts" playback of the destination, the link returns to the point of original link traversal. To reduce rewatching video, the completed destination playback returns the user to the end of the source anchor (rather than the default of returning to the point of link traversal). Having links that return to the beginning of the source anchor could help user reorientation by providing more context prior to traversal but would increase the amount of video watched multiple times.

4 Video Search User Interface

Different approaches are required to search and browse larger or less structured video collections. A typical search in a moderate to large video collection can return a large number of results. Our user interface directs the user's attention to the video segments that are potentially relevant. We present results in a form that enables users to quickly decide which of the results best satisfy the user's original information need. Our system displays search results in a highly visual form that makes it easy for users to determine which results are truly relevant.

Initially the data is divided into programs, typically 20 to 30 minutes long; and video shots, which vary from 2 to 30 seconds. Because the frames of a video shot are visually coherent, each shot can be well represented by a single keyframe. A keyframe is an image that visually represents the shot, typically chosen as a representative from the frames in the shot [35]. A time-aligned transcript obtained through automatic speech recognition (ASR) is used to provide material by which to index the shots, but because shots and their associated transcript text are typically too short to be used as effective search units, we automatically groups shots into stories and use these as the segments over which to search. Adjacent shots with relatively high text-based similarity are grouped into stories. These stories form the organizing units under which video shots are presented in our interface. By grouping related shots together we arrive at transcript text segments that are long enough to form the basis for good keyword and latent semantic indexing. Because each story comprises several shots, it cannot be well represented by a single keyframe. Instead, we represent stories as collages of the underlying shot keyframes.

Figure 6 shows the interactive search interface. The user enters a query as keywords and/or images (Fig. 6B). Keywords are typed and images are dragged into the query section from other parts of the interface. For the TRECVID interactive search task [324], the text, images, and clips describing the information need are displayed in Fig. 6C. In this case, the user can select keywords and images from the topic description. Once the user has entered a query and pressed the search button, story results appear in Fig. 6A, displayed in relevance order. The user can explore a retrieved story by clicking on the collage. The parent video is opened and the selected story is highlighted in the video timeline (Fig. 6E). Below the timeline the keyframes from all the shots in the selected story are displayed (see Fig. 6F). The shot or story under the mouse is magnified in the space in Fig. 6D. A tool tip provides additional information for the shot or story under the mouse. The user drags any shots of interest to the area shown in Fig. 6G to save relevant results. Another novel aspect of our system is that we mark visited stories so that the user can avoid needless revisiting of stories. We present the three types of UI elements that we developed:

- Three visualizations provide different information perspectives about query results.

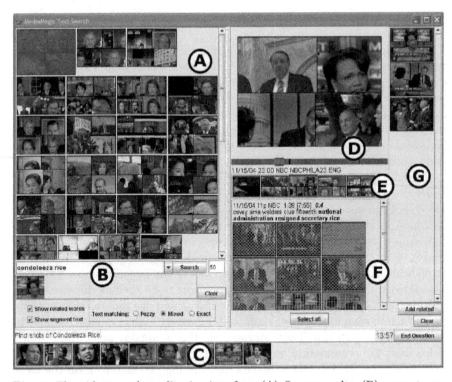

Fig. 6. The video search application interface. (A) Query results, (B) query terms and images, (C) the TRECVID topic text and example images, (D) thumbnail zoom and media player area, (E) timeline, (F) expanded shots, (G) selected shots

- Tooltips and magnified keyframes provide users with document information relevant to the query.
- Overlays provide cues about previously visited stories, current story and shot in video playback, and the degree of query relevance on story and shot.

4.1 Query Result Visualizations: Story Collage, Shot Keyframe, Video Timeline

As shown, query results are returned as a set of stories, sorted by relevance. A novel feature of our system is that retrieved stories are represented by keyframe collages where keyframes are selected and sized by their relevance to a query so that the same story may be shown differently for different queries. The size of the collage is determined by the relevance to the query so that one can see at a glance which stories are most relevant. We use a collage of four keyframes to indicate the different shots in a story without making the keyframes too small for recognizing details. We use rectangular areas for the keyframes for the sake of fast computation but we could instead use other collages such as a stained glass window visualization [59].

In addition to determining the relevance of stories with respect to the query, we also determine the relevance of each video shot. While the shot relevance does not provide good results on its own, it can be used to determine which shots within a story are the most relevant. The keyframes corresponding to the most relevant shots are combined to form a story keyframe-collage. The size allotted to each portion in this 4-image montage is determined by the shot's score relative to the query. Figure 7 shows an example of this where the query was "Condoleeza Rice" and the shots most relevant to the query are allocated more room in the story thumbnail. In this case 3 of the 4 shots selected for the story collage chosen from the original 9 shots depictCondoleeza Rice. Rather than scaling down the keyframes to form the collage, they are cropped to preserve details in reduced-size representations. In the current implementation, the top-center portion of the cropped frame is used but a sensible alternative would be to crop around the main region-of-interest as determined by color and motion analysis or face detection.

Because the automatic story segmentation is not always accurate and related stories frequently are located in the same part of the video program, we provide easy access to the temporal neighborhood of the selected story. First, the timeline of the video containing the story color-codes the relevance of all stories in the video (see Fig. 6E and Fig. 8). This color-coding provides a very distinct pattern in the case of literal text search because only few stories contain the exact keywords. After a search using the latent semantic index (LSI), all parts of the timeline indicate some relevance because every term has some latent relationship to all other terms. We experimentally determined a nonlinear mapping of the relevance scores from LSI-based text search that highlights the most related stories without completely suppressing other potentially related stories. Immediately below the timeline in Fig. 6E collages of neighboring stories around the selected story are displayed. This provides quick access to keywords in those stories via tool tips. By clicking on the timeline or the neighboring collages, the corresponding story can be expanded for closer inspection.

Fig. 7. Story keyframe collage. 9 shots shown on the left are summarized by cropped sections of the 4 shots most relevant to the query "condoleeza rice"

Fig. 8. Timelines color-coded with query relevance

Fig. 9. Tool tip showing distinguishing keywords and bold query keywords

The keyframes for the shots comprising the selected story are shown in a separate pane (see Fig. 6F and Fig. 9). Double-clicking a keyframe plays the corresponding video shot. The expanded view provides access to the individual shots for playback, for adding them to the results, and for displaying information about the shots. One or more keyframes of shots can be dragged into or out of the result area to add or remove them from the result list, or into or out of the image search area to add or remove them from the image search. Shots can also be marked explicitly as irrelevant.

4.2 Feedback on Document Relevance: Tooltips and Magnified Keyframes

It is useful to provide feedback to the user to indicate why a particular document was deemed relevant to the query and how the document is different from others in the collection. Tooltips for story collage and video shot keyframes provide that information to the user in the form of keywords that are distinctive for the story and keywords related to the query (see the plain text in Fig. 9). Terms that occur frequently in the story or shot and do not appear in many other stories or shots are most distinguishing. While words such as "lately" do not really help in distinguishing the video passage from others, words such as "russia" are helpful.

The terms in bold are most related to the query and indicate why the document is relevant to the query. We decided against displaying the terms with surrounding text as is frequently done in Web search engines. The reason is that we do not want the tooltips to be too large. Furthermore, the automatic speech recognition makes mistakes that are more noticeable when displaying whole phrases. By displaying five keywords from each of the two categories, it is likely that at least one or two are truly useful.

With a literal text search approach, the terms most related to the query are the query terms appearing in the story. When the latent semantic index is used for search, a relevant document may not contain any of the query terms but only terms that are closely related. We use the latent semantic index to identify terms in the document that are most similar to the query.

In an earlier version of our application, we displayed keyframes as part of the tooltips. Users interacting with that version of the application found that the keyframes were either too small to be useful or that the tooltips covered up too much the window. To address this issue, we reuse the video player area as a magnifier for the keyframe under the mouse or the selected keyframe (see Fig. 6D). Usually, the video player will be stopped while the user inspects keyframes so that the user can see a magnified version of the keyframe or collage without the need to dedicate some window area to that purpose.

4.3 Overlay Cues: Visited Story, Current Playback Position, Query Relevance

Semi-transparent overlays are used to provide three cues. A gray overlay on a story icon indicates that it has been previously visited (see Fig. 6A and E). A translucent red overlay on a shot icon indicates that it has been explicitly excluded by the user from the relevant shot set. A translucent green overlay on a shot icon indicates that it has been included in the results set (see Fig. 6F). Figure 9 shows the use of patterns instead of translucent overlays for color-blind users and grayscale reproduction of the image. Red diagonal lines indicate exclusion and green horizontal and vertical lines indicate inclusion.

While video is playing, the shot and the story containing the current playback position are indicated by placing a red dot on top of their keyframes. The playback position is also indicated in the timeline by a vertical red line.

Horizontal colored bars are used along the top of stories and shots to indicate the degree of query-relevance, varying from black to bright green. The same color scheme is used in the timeline depicted in Fig. 8.

5 Conclusions

In this chapter, we presented two approaches enabling users to quickly identify passages of interest within a potentially large collection of videos. The first approach automatically generates an interactive multi-level summary in the form of a hypervideo. It lets users watch short video clips and request more detail for interesting clips. Two goals for the design of hypervideo summaries are to minimize user disorientation resulting from link navigation and to minimize the rewatching of video segments. These goals are sometimes in conflict. Playing the same clip multiple times can be used to provide greater context and thus reduce disorientation. Clip selection and link generation algorithms interact in determining the degree to which a hypervideo generation approach will meet these goals.

The second approach allows users to search large video collections. We process the linear text transcript associated with the videos and segment it into stories to create three levels of segmentation (program, story, shot). Visualization techniques take advantage of the segmentation to draw the users' attention to promising query results and support them in browsing and assessing relevance. Rather than relying on elaborate media analysis techniques, we have used simple and proven automatic methods. We integrate the analysis results with an efficient user interface to enable users to both quickly browse retrieved video shots and to determine which are relevant.

Modern tools for browsing and searching video have the potential to dramatically improve access into large video libraries and to aid location of desired content within long videos. The results from Hyper-Hitchcock and our interactive video search application help establish the basis for the next generation of access tools combining searching and browsing of video.

References

1. C.B. Abdelkader and L. Davis. Detection of load-carrying people for gait and activity recognition. In *International Conference on Automatic Face and Gesture Recognition*, pages 378–383, 2002.
2. T. Adamek, N. O'Connor, and N. Murphy. Region-based segmentation of images using syntactic visual features. In *WIAMIS2005*, 2005.
3. B. Adams, C. Dorai, and S. Venkatesh. Towards automatic extraction of expressive elements from motion pictures: Tempo. In *IEEE International Conference on Multimedia and Expo*, volume II, pages 641–645, July 2000.
4. B. Adams, C. Dorai, and S. Venkatesh. Automated film rhythm extraction for scene analysis. In *IEEE International Conference on Multimedia and Expo*, pages 1056–1059, August 2001.
5. Y. Adini, Y. Moses, and S. Ullman. Face recognition: The problem of compensating for changes in illumination direction. *IEEE Transactions on Pattern Analysis and Machine Intelligence*, 19(7):721–732, 1997.
6. ADL. SCORM 2004 documentation. *Downloadable at http://www.adlnet.org/scorm/history/2004/index.cfm*, 2004.
7. J.F. Allen. Maintainning knowledge about temporal intervals. *CACM*, 26:832–843, 1983.
8. A.A. Amini, T.E. Weymouth, and R.C. Jain. Using dynamic programming for solving variational problems in vision. *IEEE Trans. Pattern Anal. Machine Intell.*, 12(9):855–867, September 1990.
9. B. Anderson and J. Moore. *Optimal Filtering*. Englewood Cliffs, NJ: Prentice-Hall, 1979.
10. O. Arandjelović and R. Cipolla. An illumination invariant face recognition system for access control using video. *In Proc. British Machine Vision Conference*, pages 537–546, 2004.
11. O. Arandjelović and R. Cipolla. An information-theoretic approach to face recognition from face motion manifolds. *Image and Vision Computing*, 2005. (to appear).
12. O. Arandjelović, G. Shakhnarovich, J. Fisher, R. Cipolla, and T. Darrell. Face recognition with image sets using manifold density divergence. *In Proc. IEEE Conference on Computer Vision and Pattern Recognition*, 1:581–588, 2005.
13. M. Ardebilian, X.W. TU, L. Chen, and P. Faudemay. Video segmentation using 3d hints contained in 2d images. *SPIE*, 2916, 1996.

14. D. Arijon. *Grammar of the film language*. Silman-James Press, 1976.
15. F. Arman, R. Depommier, A. Hsu, and M.Y. Chiu. Content-based video indexing and retrieval. In *ACM Multimedia (ACMMM'94)*, pages 97–103, San Francisco, California, 1994.
16. F. Arman, A. Hsu, and M. Chiu. Feature management for large video databases. In *In Storage and Retrieval for Image and Video Databases*, pages 2–12, 1993.
17. M.S. Arulampalam, S. Maskell, N. Gordon, and T. Clapp. A tutorial on particle filters for online non-linear/non-Gaussian Bayesian filtering. *IEEE Trans. on signal processing*, 50(2):174–188, 2002.
18. Y. Avrithis, A. Doulamis, N. Doulamis, and S. Kollias. A stochastic framework for optimal key frame extraction from mpeg video databases. *cviu*, 75(1):3–24, July 1999.
19. N. Babaguchi, Y. Kawai, and T. Kitahashi. Event based indexing of broadcast sports video by intermodal collaboration. *IEEE Transactions on Multimedia*, 4(1):68–73, 2002.
20. Y. Bar-Shalom and T.E. Fortmann. *Tracking and Data Association*. Academic Press, Orlando, Florida, 1988.
21. W.A. Barrett. A survey of face recognition algorithms and testing results. *Systems and Computers*, 1:301–305, 1998.
22. M. Basseville. Detecting changes in signals and systems - a survey. *Automatica*, 24(3):309–326, 1988.
23. Martin J. Bastiaans. Optimum sampling distances in the gabor scheme. *Proc. CSSP-97, Mierlo, Netherlands*, pages 35–42, 1997.
24. S. Belongie, C. Carson, H. Greenspan, and J. Malik. Color-and texture-based image segmentation using em and its application to content-based image retrieval. *iccv98*, pages 675–682, January 1998.
25. T.L. Berg, A.C. Berg, J. Edwards, M. Maire, R. White, Y.W. Teh, E. Learned-Miller, and D.A. Forsyth. Names and faces in the news. *In Proc. IEEE Conference on Computer Vision and Pattern Recognition*, 2004.
26. A. Beritez, S. Paek, S.-F. Chang, A. Puri, Q. Huang, J.R. Smith, C.-S Li, L.D. Bergman, and C.N. Judice. Object-based multimedia content description schemes and applications for mpeg-7. *Signal Process.: Image Commun.*, 16:235–269, 2000.
27. S.T. Birchfield. Elliptical head tracking using intensity gradients and color histograms. In *Proc. IEEE Int'l Conf. on Comput. Vis. and Patt. Recog.*, pages 232–237, 1998.
28. M. Black and A. Jepson. Eigentracking : robust matching and tracking of articulated objects using a view-based representation. *Int. Journal of Computer Vision*, 26(1):63–84, 1998.
29. M.J. Black and A.D. Jepson. Recognizing facial expressions in image sequences using local parameterized models of image motion. *International Journal of Computer Vision*, 26(1):63–84, 1998.
30. V. Blanz and T. Vetter. A morphable model for the synthesis of 3D faces. *In Proc. Conference on Computer Graphics*, pages 187–194, 1999.
31. J. Bonastre and C. Wellekens. A speaker tracking system based on speaker turn detection for NIST evaluation. *ICASSP'00*, 2000.
32. C.G. v.d. Boogaart. *Master thesis: Eine signal-adaptive Spektral-Transformation für die Zeit-Frequenz-Analyse von Audiosignalen*. Technische Universität München, 2003.

33. C.G. v.d. Boogaart and R. Lienhart. Fast gabor transformation for processing high quality audio. Technical Report 2005-21, University of Augsburg, Institute of Computer Science, 2005.
34. Y. Boon-Lock and B. Liu. A unified approach to temporal segmentation of motion jpeg and mpeg compressed video. In *IEEE International Conference on Multimedia Computing and Systems*, Los Alamitos, Ca, USA, 1995.
35. J.S. Boreczky and L.A. Rowe. Comparison of video shot boundary detection techniques. In *Storage and Retrieval for Image and Video Databases (SPIE)*, pages 170–179, 1996.
36. J.S. Boreczky and L.A. Rowe. A comparison of video shot boundary detection techniques. In *Storage and Retrieval for Image and Video Databases IV, I.K. SPIE 2670*, pages 170–179, 1996.
37. P. Bouthemy, C. Garcia, R. Ronfard, and G. Tziritas. Scene segmentation and image feature extraction for video indexing and retrieval. In *Visual Information and Information Systems*, pages 245–252, 1996.
38. P. Browne, A.F. Smeaton, N. Murphy, N. O'Connor, S. Marlow, and C. Berrut. Evaluating and combining digital video shot boundary detection algorithms. In *Irish Machine Vision and Image Processing*, 2000.
39. P. Browne, A.F. Smeaton, N. Murphy, N. O'Connor, S. Marlow, and C. Berrut. Evaluating and combining digital video shot boundary detection algorithms. In *IMVIP2000*, 2000.
40. C.J.C. Burges. A tutorial on support vector machines for pattern recognition. *Data Mining and Knowledge Discovery*, 2(2):121–167, 1998.
41. C.J.C. Burges. A tutorial on support vector machines for pattern recognition. *Data Mining and Knowledge Discovery*, 2(2):1–47, 1998.
42. M. Cai, J. Song, and M. R. Lyu. A new approach for video text detection. In *IEEE International Conference on Image Processing*, pages 117–120, 2002.
43. J. Calic and E. Izquierdo. Efficient key-frame extraction and video analysis. In *ITCC'02*, pages 28–33, 2002.
44. J. Calic and B.T. Thomas. Spatial analysis in key-frame extraction using video segmentation. In *Workshop on Image Analysis for Multimedia Interactive Services (WIAMIS'2004)*, Lisboa, Portugal, 2004.
45. C. Carson, M. Thomas, S. Belongie, J. Hellerstein, and J. Malik. Object retrieval - blobworld: A system for region-based image indexing and retrieval. *Lecture Notes in Computer Science*, 1614:509–516, 1999.
46. G. Celeux and G. Govaert. Gaussian Parsimonious Models. *Pattern Recognition*, 28(5):781–783, 1995.
47. T. Chambel, C. Zahn, and M. Finke. *Cognitively informed systems: Utilizing practical approaches to enrich information presentation and transfer*, chapter Hypervideo and cognition: Designing video based hypermedia for individual learning and collaborative knowledge building. Hershey, PA, USA: Idea Group Inc, 2005.
48. H.S. Chang, S. Sull, and S.U. Lee. Efficient video indexing scheme for content-based retrieval. *IEEE Transactions on Circuits and Systems for Video Technology*, 9(8):1269–1279, 1999.
49. P. Chang, M. Han, and Y. Gong. Extract highlights from baseball game video with hidden markov models. In *IEEE International Conference on Image Processing (ICIP'02)*, Rochester, NY, 2002.

50. S.-F. Chang, W. Chen, H. Meng, H. Sundaram, and D. Zhong. A fully auto-mated content-based video search engine supporting spatiotemporal queries. *IEEE Transactions on Circuits and Systems for Video Technology*, 1998.

51. S.-F. Chang, W. Chen, H.J. Meng, H. Sundaram, and D. Zhong. VideoQ-an automatic content-based video search system using visual cues. In *ACM Multimedia Conference*, 1997.

52. S.F. Chang and D.G. Messerschmitt. Manipulation and compositing of mc-dct compressed video. *IEEE Journal on Selected Areas of Communications*, 13(1):1–11, 1995.

53. S. Chen and P. Gopalakrishnan. Speaker, environment and channel change de-tection and clustering via the Bayesian information criterion. *Proc. of DARPA Broadcast News Transcription and Understanding Workshop*, 1998.

54. S.-C. Chen, M.-L. Shyu, C. Zhang, L. Luo, and M. Chen. Detection of soc-cer goal shots using joint multimedia features and classification rules. In *MDM/KDD*, Washington, DC, USA, August 27 2003.

55. Y. Chen, Y. Rui, and T.S. Huang. JPDAF based HMM for real-time contour tracking. In *Proc. IEEE Int'l Conf. on Comput. Vis. and Patt. Recog.*, pages I:543–550, 2001.

56. Y. Chen, Y. Rui, and T.S. Huang. Mode based multi-hypothesis head tracking using parametric contours. In *Proc. Int'l Conf. Automatic Face and Gesture Recognition*, pages 112–117, 2002.

57. Y. Chen, Y. Rui, and T.S. Huang. Parametric contour tracking using unscented Kalman filter. In *Proc. IEEE Int'l Conf. on Image Processing*, pages III: 613–616, 2002.

58. C. Chesnaud, P. Refregier, and V. Boulet. Statistical region snake-based seg-mentation adapted to different physical noise models. *IEEE Trans. Pattern Anal. Machine Intell.*, 21(11):1145–1157, November 1999.

59. P. Chiu, A. Girgensohn, and Q. Liu. Stained-Glass visualization for highly condensed video summaries. In *Proceedings of the 2004 IEEE International Conference on Multimedia and Expo, ICME 2004*, pages 2059–2062. IEEE, Jun. 2004.

60. T. Chiueh, T. Mitra, A. Neogi, and C. Yang. Zodiac: A history-based in-teractive video authoring system. *ACM/Springer-Verlag Multimedia Systems journal*, 8(3):201–211, 2000.

61. M. Christel, A. Hauptmann, H. Wactlar, and T. Ng. Collages as dynamic summaries for news video. In *ACM MM'02*, pages 561–569, 2002.

62. M. Christel, J. Yang, R. Yan, and A. Hauptmann. Carnegie mellon university search. In *TREC Video Retrieval Evaluation Online Proceedings*, 2004.

63. M.G. Christel, M.A. Smith, C.R. Taylor, and D.B. Winkler. Evolving video skims into useful multimedia abstractions. In *CHI '98: Proceedings of the SIGCHI conference on Human factors in computing systems*, pages 171–178, New York, NY, USA, 1998. ACM Press/Addison-Wesley Publishing Co.

64. T.S. Chua, S.F. Chang, L. Chaisorn, and W. Hsu. Story boundary detection in large broadcast news video archives: techniques, experience and trends. In *ACM MM'04*, pages 656–659, 2004.

65. A.J. Colmenzrez and T.S. Huang. Face detection with information-based max-imum discrimination. In *Proc. IEEE Int'l Conf. on Comput. Vis. and Patt. Recog.*, pages 782–787, 1997.

66. D. Comaniciu, V. Ramesh, and P. Meer. Real-time tracking of non-rigid objects using mean shift. In *Proc. IEEE Int'l Conf. on Comput. Vis. and Patt. Recog.*, pages II 142–149, 2000.
67. D. Comaniciu, V. Ramesh, and P. Meer. Kernel-based object tracking. IEEE *Transactions on Pattern Analysis and Machine Intelligence*, 25(5):564–577, 2003.
68. E. Cooke, P. Ferguson, G. Gaughan, C. Gurrin, G. Jones, H. Le Borgne, H. Lee, S. Marlow, K. Mc Donald, M. McHugh, N. Murphy, N. O'Connor, N. O'Hare, S. Rothwell, A.F. Smeaton, and P. Wilkins. TRECVid 2004 Experiments in Dublin City University. In *TRECVid2004*, 2004.
69. E. Cooke, P. Ferguson, G. Gaughan, C. Gurrin, G.J.F. Jones, H. Le Borgue, H. Lee, S. Marlow, K. McDonald, M. McHugh, N. Murphy, N.E. O'Connor, N. O'Hare, S. Rothwell, A.F. Smeaton, and P. Wilkins. Trecvid 2004 experiments in dublin city university. In *TREC Video Retrieval Evaluation Online Proceedings*, 2004.
70. M. Cooper. Video segmentation combining similarity analysis and classification. In *MULTIMEDIA '04: Proceedings of the 12th annual ACM international conference on Multimedia*, pages 252–255, New York, NY, USA, 2004. ACM Press.
71. M. Cooper and J. Foote. Summarizing video using non-negative similarity matrix factorization. In *IEEE Workshop on Multimedia Signal Processing (MMSP'02)*, pages 25–28, St. Thomas, US Virgin Islands, 2002.
72. D. Cristinacce, T.F. Cootes, and I. Scott. A multistage approach to facial feature detection. *In Proc. British Machine Vision Conference*, 1:277–286, 2004.
73. R. Cutler and L.S. Davis. Robust real-time periodic motion detection, analysis, and applications. *IEEE Transactions on Pattern Analysis and Machine Intelligence*, 22(8):781–796, 2000.
74. D. Frey D, R. Gupta, V. Khandelwal, V. Lavrenko, A. Leuski, and J. Allan. Monitoring the news: a TDT demonstration system. In *HLT'01*, pages 351–355, 2001.
75. Ingrid Daubechies. *Ten lectures on wavelets.* Society for Industrial and Applied Mathematics, Philadelphia, PA, USA, 1992.
76. M. Davis. Knowledge representation for video. *Twelfth Conference on Artificial Intelligence*, pages 120–127, 1994.
77. D. DeMenthon, V. Kobla, and D. Doermann. Video summarization by curve simplification. In *ACM International Conference on Multimedia*, pages 211–218, 1998.
78. A.P. Dempster, N.M. Laird, and D.B. Rubin. Maximum likelihood from incomplete data (with discussion). *Journal of the Royal Statistical Society, Series B*, 39:1–38, 1977.
79. Y. Deng and B.S. Manjunath. Netra-v: Toward an object-based video representation. *IEEE Transactions on Circuits and Systems for Video Technology*, 8(5):616–627, 1998.
80. J. Denzler and H. Niemann. Active-rays: Polar-transformed active contours for real-time contour tracking. *Real-Time Imaging*, 5(3):203–13, June 1999.
81. V. Di Lecce, A. Dimauro, G. and Guerriero, S. Impedovo, G. Pirlo, and A. Salzo. Image basic features indexing techniques for video skimming. In *IEEE International Conference on Image Analysis and Processing*, pages 715–720, 1999.

82. C. Dickie, R. Vertegaal, D. Fono, C. Sohn, D. Chen, D. Cheng, J. Shell, and O. Aoudeh. Augmenting and sharing memory with eyeblog. In *The First ACM Workshop on Continuous Archival and Retrieval of Personal Experiences*, 2004.

83. D. Diklic, D. Petkovic, and R. Danielson. Automatic extraction of representative keyframes based on scene content. In *Conference Record of the Thirty Second Asilomar Conference on Signals, Systems Computers*, pages 877–881, 1998.

84. N. Dimitrova, H-J Zhang, B. Shahraray, I. Sezan, T. Huang, and A. Zakhor. Applications of video-content analysis and retrieval. *IEEE Multimedia*, 9(3):42–55, 2002.

85. W. Ding, G. Marchionini, and T. Tse. Previewing video data: browsing key frames at high rates using a video slide show interface. In *International Symposium on Research, Development and Practice in Digital Librairies (ISDL'97)*, pages 425–426, 1997.

86. D. Donoho. Nonlinear wavelet methods for recovery of signals, densities, and spectra from indirect and noisy data. In *Different Perspectives on Wavelets*, volume 47 of *Proceeding of Symposia in Applied Mathematics*, pages 173–205, 1993.

87. Chitra Dorai and Svetha Venkatesh. Computational Media Aesthetics: Finding meaning beautiful. *IEEE Multimedia*, 8(4):10–12, October-December 2001.

88. F. Dufaux. Key frame selection to represent a video. In *International onference on Image Processing*, pages 275–278, 2000.

89. J.P. Eakins. Towards intelligent image retrieval. *Pattern Recognition*, 35(1):314, 2002.

90. G.J. Edwards, C.J. Taylor, and T.F. Cootes. Interpreting face images using active appearance models. *IEEE International Conference on Automatic Face and Gesture Recognition*, pages 300–305, 1998.

91. A. Ekin and A.M. Tekalp. A framework for analysis and tracking of soccer video. In *SPIE/IS&T Conference on Visual Communication Image Processing*, 2002.

92. http://www.ephyx.com/.

93. B. Erol, J. Hull, and D. Lee. Linking multimedia presentations with their symbolic source documents: Algorithm and applications. *ACM Multimedia*, pages 498–507, 2003.

94. M. Everingham and A. Zisserman. Automated person identification in video. *International Conference on Image and Video Retrieval*, pages 289–298, 2004.

95. R. Fablet, P. Bouthemy, and P. Perez. Non-parametric motion characterization using causal probabilistic models for video indexing and retrieval. *IEEE Trans. on Image Processing*, 11(4), apr 2002.

96. J. Fan, A.K. Elmagarmid, X. Zhu, W.G. Aref, and Lide Wu. Classview: Hierarchical video shot classification, indexing, and accessing. *IEEE Transactions on Multimedia*, 6(1), February 2004.

97. Hans G. Feichtinger and Thomas Strohmer. *Gabor Analysis and Algorithms: Theory and Applications*. Birkhäuser, Boston, 1998.

98. Hans G. Feichtinger and Thomas Strohmer. *Advances in Gabor Analysis*. Birkhäuser, Boston, 2003.

99. P.F. Felzenszwalb and D. Huttenlocher. Pictorial structures for object recognition. *International Journal of Computer Vision*, 61(1):55–79, 2005.

100. A.M. Ferman and Murat Tekalp. Multiscale content extraction and representation for video indexing. In *SPIE on Multimedia Storage and Archiving Systems II*, pages 23–31, 1997.

101. A. Fern, R. Givan, and J.M. Siskind. Specific-to-general learning for temporal events with applications to learning event definitions from video. *Journal of Artificial Intelligence Research*, 17:379–449, 2002.

102. G.D. Finlayson and G.Y. Tian. Color normalization for color object recognition. *International Journal of Pattern Recognition and Artificial Intelligence*, 13(8):1271–1285, 1999.

103. A. Fitzgibbon and A. Zisserman. On affine invariant clustering and automatic cast listing in movies. *In Proc. IEEE European Conference on Computer Vision*, pages 304–320, 2002.

104. M. Flickner, H. Sawhney, W. Niblack, J. Ashley, Q. Huang, B. Dom, M. Gorkani, J. Hafner, D. Lee, D. Petkovic, D. Steele, and P.Yanker. Query by image and video content: The qbic system. *IEEE Computer*, 38:23–31, 1995.

105. K. Fukui and O. Yamaguchi. Facial feature point extraction method based on combination of shape extraction and pattern matching. *Systems and Computers in Japan*, 29(6):2170–2177, 1998.

106. K. Fukunaga. *Introduction to Statistical Pattern Recognition*. Academic Press, 1990.

107. Dennis Gabor. Theory of communication. *Journal of the Institution of Electrical Engineers*, pages 429–457, November 1946.

108. A. Galata, N. Johnson, and D. Hogg. Learning variable-length markov models of behavior. *Computer Vision and Image Understanding*, 81(3):398–413, 2001.

109. D. Geiger, A. Gupta, L.A. Costa, and J. Vlontzos. Dynamic-programming for detecting, tracking, and matching deformable contours. *IEEE Trans. Pattern Anal. Machine Intell.*, 17(3):294–302, 1995.

110. M. Gelgon, P. Bouthemy, and T. Dubois. A region tracking method with failure detection for an interactive video indexing environment. In *Proc of Int. Conf on Visual Information Systems (Visual'99)*, pages 261–268, Amsterdam, June 1999. LNCS 1614.

111. J. Gemmell, L. Williams, G. Wood, K. and Bell, and R. Lueder. Passive capture and ensuing issues for a personal lifetime store. In *The First ACM Workshop on Continuous Archival and Retrieval of Personal Experiences*, 2004.

112. T. Gevers and A. Smeulders. Image indexing using composite color and shape invariant features. In *Proceedings of the 6th International Conference on Computer Vision, Bombay, India*, pages 576–581, January 1998.

113. A. Girgensohn, J. Boreczky, P. Chiu, J. Doherty, J. Foote, G. Golovchinsky, S. Uchihashi, and L. Wilcox. A semi-automatic approach to home video editing. In *UIST '00: Proceedings of the 13th annual ACM symposium on User interface software and technology*, pages 81–89, New York, NY, USA, 2000. ACM Press.

114. A. Girgensohn, L. Wilcox, F. Shipman, and S. Bly. Designing affordances for the navigation of detail-on-demand hypervideo. In *AVI '04: Proceedings of the working conference on Advanced visual interfaces*, pages 290–297, New York, NY, USA, 2004. ACM Press.

115. Y. Gong. Summarizing audio-visual contents of a video program. *EURASIP J. on Applied Signal Processing: Special Issue on Unstructured Information Management from Multimedia Data Sources*, 2003(3), 2003.

116. Y. Gong and X. Liu. Video summarization and retrieval using singular value decomposition. *ACM Multimedia Systems Journal*, 9:157–168, 2003.

117. Google video search. http://video.google.com.

118. http://video.google.com.

119. G.R. Grimmett and D.R. Stirzaker. *Probability and Random Processes.* Clarendon Press, Oxford, 2nd edition, 1992.

120. W. Grimson, L. Lee, R. Romano, and C. Stauffer. Using adaptive tracking to classify and monitor activities in a site. In *Proc. of IEEE Int. Conf. on Computer Vision and Pattern Recognition (CVPR'1998)*, Santa Barbara, CA., USA, June 1998.

121. N. Guimãres, T. Chambel, and J. Bidarra. From cognitive maps to hyper-video: Supporting flexible and rich learner-centred environments. *Interactive Multimedia Journal of Computer-Enhanced Learning*, 2((2)), 2000. http://imej.wfu.edu/articles/2000/2/03/.

122. N. Haering, R. Qian, and M.I. Sezan. A semantic event detection approach and its application to detecting hunts in wildlife video. *IEEE Transactions on Circuits and Systems for Video Technology*, 10(9):857–868, 2000.

123. J. Haitsma and T. Kalker. A highly robust audio fingerprinting system. *Third International Conference on Music Information Retrieval*, pages 107–115, 2002.

124. R. Hammoud. *Building and Browsing of Interactive Videos.* PhD thesis, INRIA, Grenoble, Feb 2001. http://www.inrialpes.fr/movi/publi/Publications/2001/Ham01/index.html.

125. R. Hammoud and L. Chen. A spatiotemporal approach for semantic video macro-segmentation. In *European Workshop on Content-Based Multimedia Indexing*, pages 195–201, IRIT-Toulouse FRANCE, Octobre 1999.

126. R. Hammoud, L. Chen, and F. Fontaine. An extensible spatial-temporal model for semantic video segmentation. In *First Int. Forum on Multimedia and Image Processing, Anchorage, Alaska*, May 1998.

127. R. Hammoud and D. G. Kouam. A mixed classification approach of shots for constructing scene structure for movie films. In *Irish Machine Vision and Image Processing Conference*, pages 223–230, The Queen's University of Belfast, Northern Ireland, 2000.

128. R. Hammoud and R. Mohr. Building and browsing hyper-videos: a content variation modeling solution. *Pattern Analysis and Applications*, 2000.

129. R. Hammoud and R. Mohr. Gaussian mixture densities for indexing of localized objects in a video sequence. Technical report, INRIA, March 2000. http://www.inria.fr/RRRT/RR-3905.html.

130. R. Hammoud and R. Mohr. Interactive tools for constructing and browsing structures for movie films. In *ACM Multimedia*, pages 497–498, Los Angeles, California, USA, October 30 - November 3 2000. (demo session).

131. R. Hammoud and R. Mohr. Mixture densities for video objects recognition. In *International Conference on Pattern Recognition*, volume 2, pages 71–75, Barcelona, Spain, 3-8 September 2000.

132. R. Hammoud and R. Mohr. A probabilistic framework of selecting effective key frames for video browsing and indexing. In *International workshop on Real-Time Image Sequence Analysis*, pages 79–88, Oulu, Finland, Aug. 31-Sep. 1 2000.

133. R.I. Hammoud, A. Wilhelm, P. Malawey, and G.J. Witt. Efficient real-time algorithms for eye state and head pose tracking in advanced driver support systems. In *IEEE Computer Vision and Pattern Recognition Conference*, San Diego, CA, USA, June 20-26 2005.

134. A. Hampapur, R. Jain, and T.E. Weymouth. Production model based digital video segmentation. In *Multimedia Tools and Applications*, pages 9–46, 1995.
135. B. Han, D. Comaniciu, Y. Zhu, and L. Davis. Incremental density approximation and kernel-based Bayesian filtering for object tracking. In *IEEE Int. Conf on Computer Vision and Pattern Recognition (CVPR'2004)*, pages 638–644, Washington, USA, June 2004.
136. D.W. Hansen, R.I. Hammoud, R. Satria, and J. Sorensen. Improved likelihood function in particle-based ir eye tracking. In *IEEE CVPR Workshop on Object Tracking and Classification Beyond the Visible Spectrum*, San Diego, CA, USA, June 20, 2005.
137. J. Hartung, B. Elpelt, and K. Klösener. *Statistik. Lehr- und Handbuch der angewandten Statistik*. Oldenbourg, München, 1995.
138. A. Haubold and J. Kender. Analysis and interface for instructional video. *ICME'03*, 2003.
139. A. Hauptmann, M-Y. Chen, M. Christel, C. Huang, W-H. Lin, T. Ng, N. Papernick, A. Velivelli, J. Yang, R. Yan, H. Yang, and H. Wactlar. Confounded Expectations: Informedia at TRECVid 2004. In *TRECVid2004*, 2004.
140. L. He, E. Sanocki, A. Gupta, and J. Grudin. Auto-summarization of audio-video presentations. In *ACM Multimedia (ACMMM'99)*, pages 489–498, Orlando, Florida, 1999.
141. D. Heesch, P. Howarth, J. Megalhaes, A. May, M. Pickering, A. Yavlinsky, and S. Ruger. Video retrieval using search and browsing. In *TREC Video Retrieval Evaluation Online Proceedings*, 2004.
142. Lewis Herman. *Educational Films: Writing, Directing, and Producing for Classroom, Television, and Industry*. Crown Publishers, INC., New York, 1965.
143. K. Hirata, Y. Hara, H. Takano, and S. Kawasaki. Content-oriented integration in hypermedia systems. In *HYPERTEXT '96: Proceedings of the the seventh ACM conference on Hypertext*, pages 11–21, New York, NY, USA, 1996. ACM Press.
144. E. Hjelmås. Face detection: A survey. *Computer Vision and Image Understanding*, (83):236–274, 2001.
145. Tilman Horn. Image processing of speech with auditory magnitude spectrograms. *Acta Acustica united with Acustica*, 84(1):175–177, 1998.
146. Eva Hörster, Rainer Lienhart, Wolfgang Kellerman, and J.-Y. Bouguet. Calibration of visual sensors and actuators in distributed computing platforms. *3rd ACM International Workshop on Video Surveillance & Sensor Networks*, November 2005.
147. J. Huang. *Color Spatial Indexing and Applications*. PhD thesis, Cornell University, August 1998.
148. M. Huang, A. Mahajan, and D. DeMenthon. Automatic performance evaluation for video summarization. Technical Report LAMP-TR-114, CAR-TR-998, CS-TR-4605, UMIACS-TR-2004-47, University of Maryland, July 2001.
149. Barbara Burke Hubbard. *The World According to Wavelets: The Story of a Mathematical Technique in the Making*. A K Peters, Wellesley, Massachusetts, 1996.
150. A. Humrapur, A. Gupta, B. Horowitz, C.F. Shu, C. Fuller, J. Bach, M. Gorkani, and R. Jain. Virage video engine. In *Proc. SPIE, Storage and Retrieval for Image and Video Databases V*, February.
151. http://www.hyperfilm.it/eng/index.html.

152. IBM. MARVEL: MPEG-7 Multimedia Search Engine. website available at: http://www.research.ibm.com/marvel/ (last visited august 2005).

153. I. Ide, H. Mo, N. Katayama, and S. Satoh. Topic threading for structuring a large-scale news video archive. In *CIVR2004 (LNCS3115)*, pages 123–131, 2004.

154. IEEE-SA. Draft standard for learning object metadata. *http://ltsc.ieee.org/wg12/index.html*, 2002.

155. Informedia. Informedia digital video understanding research. website available at: http://www.informedia.cs.cmu.edu/ (last visited august 2005).

156. Intel Open Source Computer Vision, http://www.intel.com/research/mrl/research/opencv/.

157. http://www.intervideo.com/.

158. http://news.bbc.co.uk/1/hi/technology/4336194.stm.

159. M. Isard and A. Blake. Contour tracking by stochastic propagation of conditional density. In *Proc. European Conf. on Computer Vision*, pages I:343–356, 1996.

160. M. Isard and A. Blake. CONDENSATION - conditional density propagation for visual tracking. *International Journal of Computer Vision*, 1(29):5–28, 1998.

161. M. Isard and A. Blake. CONDENSATION - conditional density propagation for visual tracking. *Int. J. Computer Vision*, 29(1):5–28, 1998.

162. M. Isard and A. Blake. ICONDENSATION: Unifying low-level and high-level tracking in a stochastic framework. In *Proc. European Conf. on Computer Vision*, pages 767–781, 1998.

163. http://news.bbc.co.uk/1/hi/technology/4639880.stm.

164. M. Izumi and A. Kojima. Generating natural language description of human behavior from video images. In *International Conference on Pattern Recognition*, pages 728–731, 2000.

165. J. Sivic J and A. Zisserman. Video Google: a text retrieval approach to object matching in videos. In *ICCV 2003*, 2003.

166. A. Jacquot, P. Sturm, and O. Ruch. Adaptive tracking of non-rigid object based on histograms and automatic parameter selection. In *IEEE Workshop on motion and video computing*, pages 103–109, Breckenridge, Col., USA, January 2005.

167. A.K. Jain and S.B. Acharjee. Text segmentation using gabor filter for automatic document processing. *Machine Vision and Application*, 5(3):169–184, 1992.

168. A.K. Jain and Y. Zhong. Page segmentation in images and video frames. *Pattern Recognition*, 1998.

169. A.K. Jain and R.C. Dubes. *Algorithms for Clustering Data*. Prentice Hall, 1988.

170. Ramesh Jain and A. Hampapur. Metadata in video databases. *SIGMOD Record (ACM Special Interest Group on Management of Data)*, 23(4):27–33, 1994.

171. Z. Jing and R. Mariani. Glasses detection and extraction by deformable contour. *In Proc. International Conference on Pattern Recognition*, 2, 2000.

172. T. Joachims. *Making large-scale SVM learning practical*. MIT Press, In Advances in Kernel Methods - Support Vector Learning, 1999.

173. S.C. Johnson. Hierarchical clustering schemes. *Psychometrika*, (32):241–254, 1967.

174. N. Jojic, M. Turk, and T.S. Huang. Tracking articulated objects in dense disparity maps. In *Proc. IEEE Int'l Conf. on Computer Vision*, pages 123–130, 1999.

175. M. Jordan. Graphical models. *Statistical Science*, 19:140–155, 2004.

176. S.J. Julier and J.K. Uhlmann. A general method for approximating nonlinear transformations of probability distributions. Technical Report RRG, Dept. of Engineering Science, University of Oxford, 1996.

177. S.J. Julier and J.K. Uhlmann. Unscented filtering and nonlinear estimation. *Proceeding of the IEEE*, 92(3):401–422, 2004.

178. K. McDonald K and A.F. Smeaton. A comparison of score, rank and probability-based fusion methods for video shot retrieval. In *CIVR2005 (LNCS3569)*, pages 61–70, 2005.

179. J. Kang, I. Cohen, G. Medioni, and C. Yuang. Detection and tracking of moving objects from a mobile platform in presence of strong parallax. In *Proc. of IEEE Int. Conf. on Computer Vision (ICCV'2005)*, Beijing, China, October 2005.

180. M. Kass, A. Witkin, and D. Terzopoulos. Snakes : active contour models. In *IEEE Int. Conf. on Computer Vision (ICCV'87)*, pages 261–268, London, 1987.

181. M. Kass, A. Witkin, and D. Terzopoulos. Snakes: Active contour models. *IJCV*, 1(4):321–331, 1988.

182. T. Kawashima, K. Tateyama, T. Iijima, and Y. Aoki. Indexing of baseball telecast for content-based video retrieval. In *IEEE International Conference on Image Processing*, pages 871–874, 1998.

183. B. Kepenekci. *Face Recognition Using Gabor Wavelet Transform*. PhD thesis, The Middle East Technical University, 2001.

184. M. Kim, J.G. Jeon, J.S. Kwak, M.H. Lee, and C. Ahn. Moving object segmentation in video sequences by user interaction and automatic object tracking. *Image and Vision Computing*, 19(24):52–60, 2001.

185. D. Kimber and L. Wilcox. Acoustic segmentation for audio browsers. *Proc. of Interface Conference, Sydney, Australia*, July 1996.

186. A. Kojima, T. Tamura, and K. Fukunaga. Natural language description of human activities from video images based on concept hierarchy of actions. *International Journal of Computer Vision*, 50(2):171–184, 2002.

187. A. Kong, J.S. Liu, and W.H. Wong. Sequential imputations and Bayesian missing data problems. *Journal of the American Statistical Association*, 89:278–288, 1994.

188. W. Kraaij, A.F. Smeaton, and P. Over. TRECVid2004 - an overview. In *TRECVid2004*, 2004.

189. E. Kraft. A quaternion-based unscented kalman filter for orientation tracking. In *Proc. IEEE Int'l Conf. on Information Fusion*, pages 47–54, 2003.

190. S. Kullback. *Information theory and Statistics*. Dover, New York, NY, 1968.

191. M.P. Kumar, P.H.S. Torr, and A. Zisserman. OBJ CUT. In *IEEE Int. Conf. on Computer Vision and Pattern Recognition (CVPR'2005)*, San Diego, USA, June 2005.

192. S.S. Kuo and O.E. Agazzi. Keyword spotting in poorly printed documents using pseudo-2D hidden Markov models. *IEEE Trans. Pattern Anal. Machine Intell.*, 16(8):842–848, August 1994.

193. Joserph Laviola. A comparison of unscented and extended kalman filtering for estimating quaternion motion. In *Proc. of American Control Conference*, pages 2435–2440, June 2003.

194. H. Lee, A.F. Smeaton, N. O'Connor, and N. Murphy. User-interface to a CCTV video search system. In *IEE ICDP2005*, pages 39–44, 2005.

195. H. Lee, A.F. Smeaton, and B. Smyth. User evaluation outisde the lab: the trial of Físchlár-News. In *CoLIS5 Workshop on Evaluating User Studies in Information Access*, 2005.

196. B. Li and R. Chellappa. A generic approach to simultaneous tracking and verification in video. *IEEE Transactions on Image Processing*, 11(5):530–544, 2002.

197. B. Li, J. Errico, M. Ferman, H. Pan, P. Van Beek, and I. Sezan. Sports program summarization. In *IEEE Conference on Computer Vision and Pattern Recognition (Demonstrations)*, 2001.

198. B. Li, J. Errico, H. Pan, and I. Sezan. Bridging the semantic gap in sports video retrieval and summarization. *Journal of Visual Communication and Image Representation*, 15:393–424, 2004.

199. B. Li and I. Sezan. Event detection and summarization in sports video. In *IEEE Workshop on Content-Based Access to Video and Image Libraries*, pages 131–136, 2001.

200. B. Li and I. Sezan. Event detection and summarization in american football. In *SPIE/IS&T Conference on Visual Communication Image Processing*, pages 202–213, 2002.

201. B. Li and I. Sezan. Semantic sports video analysis and new applications. In *IEEE International Conference on Image Processing*, pages 17–20, 2003.

202. S.Z. Li, Q.D. Fu, L. Gu, B. Scholkopf, Y. Cheng, and H.J. Zhang. Kernel machine based learning for multi-view face detection and pose estimation. In *Proc. IEEE Int'l Conf. on Computer Vision*, pages II: 674–679, 2001.

203. Y. Li and C. Dorai. Video frame classification for learning media content understanding. In *Proc. SPIE ITCOM Conference on Internet Multimedia Management Systems V*, Philadelphia, October 2004.

204. Y. Li, S. Narayanan, and C.-C. Kuo. Identification of speakers in movie dialogs using audiovisual cues. *ICASSP'02*, Orlando, May 2002.

205. Ying Li, Shrikanth Narayanan, and C.C. Jay Kuo. *Video Mining*, chapter Chapter 5: Movie content analysis, indexing and skimming via multimodal information. Kluwer Academic Publishers, 2003.

206. R. Lienhart. Dynamic video summarization of home video. In *Storage and Retrieval for Media Databases*, volume 3972 of *Proceedings of SPIE*, pages 378–389, 2000.

207. R. Lienhart, S. Pfeiffer, and W. Effelsberg. Video abstracting. *Communications of ACM*, pages 55–62, 1997.

208. R. Lienhart, S. Pfeiffer, and W. Effelsberg. Scene determination based on video and audio features. In *IEEE Conference on Multimedia Computing and Systems*, pages 07–11, 1999.

209. R. Lienhart and F. Stuber. Automatic text recognition in digital videos. In *Proc. Praktische Informatic IV*, pages 68–131, 1996.

210. R. Lienhart and A.Wernicke. Localizing and segmenting text in images, videos and web pages. *IEEE Transactions on Circuits and Systems for Video Technology*, 12(4):256–268, April 2002.

211. C-Y. Lin, B.L. Tseng, and J. Smith. Video collaborative annotation forum: establishing ground-truth labels on large multimedia datasets. In *TRECVid2003*, 2003.

212. T. Liu, H.J. Zhang, and F. Qi. A novel video key-frame extraction algorithm based on perceived motion energy model. *IEEE Transaction on Circuits and Systems for Video Technology*, 13(10):1006–1013, 2003.

213. Z. Liu and Q. Huang. Detecting news reporting using av information. In *IEEE International Conference on Image Processing*, 1999.

214. L. Lu, S.Z. Li, and H.J. Zhang. Content-based audio segmentation using support vector machine. *ICME'01*, 2001.

215. L. Lu, H. Zhang, and H. Jiang. Content analysis for audio classification and segmentation. *IEEE Transactions on Speech and Audio Processing*, 10(7), 2002.

216. Bruce D. Lucas and Takeo Kanade. An iterative image registration technique with an application to stereo vision. In *Proceedings of the 7th International Joint Conference on Artificial Intelligence (IJCAI '81)*, pages 674–679, April 1981.

217. H. Luo and A. Eleftheriadis. Designing an interactive tool for video object segmentation and annotation. In *ACM Int. Conf. on Multimedia*, pages 265–269, Orlando, Florida, USA, November 1999.

218. Henrique S. Malvar. *Signal Processing with Lapped Transforms*. Artech House, Inc., Norwood, MA, USA, 1992.

219. I. Mani and M.T. Maybury. *Advances in Automatic Text Summarization*. The MIT Press, 1999.

220. B.S. Manjunath, P. Salembier, T. Sikora, and P. Salembier. *Introduction to MPEG 7: Multimedia Content Description Language*. John Wiley & Sons, New York, June 2002.

221. G. Marchionini and G. Geisler. The open video digital library. *D-Lib Magazine*, 8(12), 2002.

222. V.Y. Mariano and R. Kasturi. Locating uniform-colored text in video frames. In *International Conference on Pattern Recognition*, volume 4, pages 539–542, 2000.

223. A.M. Martinez. Recognizing imprecisely localized, partially occluded and expression variant faces from a single sample per class. *IEEE Transactions on Pattern Analysis and Machine Intelligence*, 24(6):748–763, 2002.

224. S.J. Mckenna, S. Gong, and Y. Raja. Modelling facial colour and identity with Gaussian mixtures. *Pattern Recognition*, 31(12):1883–1892, 1998.

225. G. Mekenkamp, M. Barbieri, B. Huet, I. Yahiaoui, B. Merialdo, and R. Leonardi. Generating tv summaries for ce-devices. In *MM'02, 10th International ACM Conference on Multimedia*, pages 83–84, Juan-les-Pins, France, December 1-6 2002.

226. K. Mikolajczyk, R. Choudhury, and C. Schmid. Face detection in a video sequence - a temporal approach. *In Proc. IEEE Conference on Computer Vision and Pattern Recognition*, 2:96–101, 2001.

227. http://today.reuters.com/news/NewsArticle.aspx?type=technologyNews& storyID=2006-01-23T143929Z_01_HEL003549_RTRUKOC_0_US-NOKIA-MOBILE-TV.xml.

228. K. Mori and S. Nakagawa. Speaker change detection and speaker clustering using VQ distortion for broadcast news speech recognition. *ICASSP'01*, 1:413–416, May 2001.

229. http://www.artsvideo.com/movideo/.

230. B.A. Myers, J.P. Casares, S. Stevens, L. Dabbish, D. Yocum, and A. Corbett. A multi-view intelligent editor for digital video libraries. In *JCDL '01: Proceedings of the 1st ACM/IEEE-CS joint conference on Digital libraries*, pages 106–115, New York, NY, USA, 2001. ACM Press.

231. P. Mylonas, K. Karpouzis, G. Andreou, and S. Kollias. Towards an integrated personalized interactive video. In *IEEE Symp. on multimedia software engineering*, Miami, Florida, December 2004.

232. F. Nack. All content counts: the future of digital media is meta. *IEEE Multimedia*, 7:10–13, July-September, 2000.

233. F. Nack and A.T. Lindsay. Everything you wanted to know about mpeg-7. *IEEE Multimedia*, pages 65–77, July-September 1999.

234. J. Nam and A.H. Tewfik. Video abstract of video. In *IEEE 3rd Workshop on Multimedia Signal Processing*, pages 117–122, 1999.

235. A. Nefian and M. Hayes, III. Maximum likelihood training of the embedded HMM for face detection and recognition. In *Proc. IEEE Int'l Conf. on Image Processing*, pages 33–36, 2000.

236. C. Ngo, Y. Ma, and H. Zhang. Automatic video summarization by graph modeling. In *International Conference on Computer Vision (ICCV'03)*, volume 1, Nice, France., 2003.

237. B.C. O'Connor. Selecting key frames of moving image documents. *Microcomputers for Information Management*, 8(2):119–133, 1991.

238. B.C. O'Connor. Selecting key frames of moving image documents : A digital environment for analysis and navigation. *Microcomputers for Information Management*, 8(2):119–133, 1991.

239. J-M. Odobez and P. Bouthemy. Robust multiresolution estimation of parametric motion models. *Journal of Visual Commmunication and Image Representation*, 6(4):348–365, December 1995.

240. J-M. Odobez and P.P. Bouthemy. Separation of moving regions from background in an image sequence acquired with a mobile camera. In H.H. Li, S. Sun, and H. Derin, editors, *Video Data Compression for Multimedia Computing*, pages 283–311. Kluwer Academic Publisher, 1997.

241. N. O'Hare, A.F. Smeaton, C. Czirjek, N. O'Connor, and N. Murphy. A generic news story segmentation system and its evaluation. In *ICASSP2004*, 2004.

242. N.M. Oliver, B. Rosario, and A.P. Pentland. A bayesian computer vision system for modeling human interactions. *IEEE Transations on Pattern Analysis and Machine Intelligence*, 22(8):831–843, 2000.

243. N. Omoigui, L. He, A. Gupta, J. Grudin, and E. Sanocki. Time-compression: System concerns, usage, and benefits. In *ACM Conference on Computer Human Interaction*, 1999.

244. Alan V. Oppenheim and Ronald W. Schafer. *Discrete-time signal processing*. Prentice-Hall, Inc., Upper Saddle River, NJ, USA, 1989.

245. D. O'Sullivan, B. Smyth, D. Wilson, K. McDonald, and A.F. Smeaton. Improving the quality of the personalized electronic program guide. *User Modeling and User-Adapted Interaction*, 14(1):5–36, 2004.

246. C. O'Toole, A.F. Smeaton, N. Murphy, and S. Marlow. Evaluation of automatic shot boundary detection on a large video test suite. In *Challenges for Image Retrieval*, Brighton, 1999.

247. N. Paragios and R. Deriche. Geodesic active regions and level set methods for motion estimation and tracking. *Computer Vision and Image Understanding*, (97):259–282, 2005.

248. K. Peker and A. Divakaran. Adaptive fast playback-based video skimming using a compressed-domain visual complexity measure. In *International Conference on Multimedia and Expo (ICME'04)*, 2004.

249. P. Perez, C. Hue, J. Vermaak, and M. Gangnet. Color-based probabilistic tracking. In *Proc. of European Conference on Computer Vision (ECCV'2002)*, LNCS 2350, pages 661–675, Copenhaguen, Denmark, May 2002.

250. N. Peterfreund. Robust tracking of position and velocity with Kalman snakes. *IEEE Trans. Pattern Anal. Machine Intell.*, 21(6):564–569, June 1999.

251. S. Pfeiffer, R. Lienhart, S. Fischer, and W. Effelsberg. Abstracting digital movies automatically. *Journal of Visual Communication and Image Representation*, 7(4):345–353, 1996.

252. P.J. Phillips, P.J. Flynn, T. Schruggs, K.W. Bowyer, J. Chang, K. Hoffman, J. Marques, J. Min, and W. Worek. Overview of the face recognition grand challenge. *In Proc. IEEE Conference on Computer Vision and Pattern Recognition*, 1:947–954, 2005.

253. P.J. Phillips, P. Grother, R.J. Micheals, D.M. Blackburn, E. Tabassi, and J.M. Bone. FRVT 2002: Overview and summary. *Technical report, National Institute of Justice*, 2003.

254. R. Pless, T. Brodsky, and Y. Aloimonos. Detecting independent motion : the statistics of temporal continuity. IEEE *Transactions on Pattern Analysis and Machine Intelligence*, 22(8):768–777, 2000.

255. A. Pope, R. Kumar, H. Sawhney, and C. Wan. Video abstraction: summarizing video content for retrieval and visualization. In *Conference Record of the Thirty Second Asilomar Conference*, pages 915–919, 1998.

256. A. Prati, I. Mikic, M. Trivedi, and R. Cucchiara. Detecting moving shadows: Algorithms and evaluation. *IEEE Trans. on Pattern Analysis and Machine Intelligence*, 2003.

257. S. Qian and D. Chen. *Joint time-frequency analysis: methods and applications.* Prentice Hall, New Jersey, 1996.

258. L.R. Rabiner and B.H. Juang. An introduction to hidden Markov models. *IEEE Trans. Acoust., Speech, Signal Processing*, 3(1):4–15, January 1986.

259. L.R. Rabiner. A tutorial on hidden markov models and selected applications in speech recognition. *Proceedings of the IEEE*, 77:257–285, 1989.

260. C. Rasmussen and G. Hager. Joint probabilistic techniques for tracking multi-part objects. In *Proc. IEEE Int'l Conf. on Comput. Vis. and Patt. Recog.*, pages 16–21, 1998.

261. M. Rautiainen and D. Doermann. Temporal color correlograms for video retrieval. In *ICPR2002*, pages 267–270, 2002.

262. E. Rennison. Galaxy of news: an approach to visualizing and understanding expansive news landscapes. In *UIST'94*, pages 3–12, 1994.

263. D.A. Reynolds. An overview of automatic speaker recognition technology. In *Proc. ICASSP*, volume 4, pages 4072–4075, 2002.

264. A. Ribert, A. Ennaji, and Y. Lecourtier. A multi-scale clustering algorithm. *Vision interface*, pages 592–597, may 1999.

265. G. Rigoll, S. Muller, and B. Winterstein. Robust person tracking with non-stationary background using a combined pseudo-2D-HMM and Kalman-filter approach. In *Proc. IEEE Int'l Conf. on Image Processing*, pages 242–246, 1999.

266. O. Rioul and M. Vetterli. Wavelets and signal processing. *IEEE Signal Processing Magazine*, 8(4):14–38, 1991.

267. R. Ronfard. Reading movies - an integrated DVD player for browsing movies and their scripts. In *ACM MM'04*, pages 740–741, 2004.

268. R. Rosales. Recognition of human action using moment-based features. Technical Report 98-020, Boston University Computer Science, Boston, MA, November 1998.

269. C. Rother, V. Kolmogorov, and A. Blake. Grabcut - interactive foreground extraction using iterated graph cuts. *ACM Trans. on Graphics/SIGGRAPH'2004*, 23(3):309–314, August 2004.

270. L.A. Rowe and J.M. Gonzalez. BMRC lecture browser demo. *http://bmrc.berkeley.edu/frame/projects/lb/index.html*, 2000.

271. H. Rowley, S. Baluja, and T. Kanade. Neural network-based face detection. *IEEE Transactions on Pattern Analysis and Machine Intelligence*, 20(1), 1998.

272. Y. Rui, A. Gupta, and Acero A. Automatically extracting highlights for tv baseball programs. In *ACM Multimedia*, 2000.

273. Y. Rui, A. Gupta, J. Grudin, and L. He. Automating lecture capture and broadcast: Technology and videography. *Multimedia Systems*, 10(1):3–15, june 2004.

274. Y. Rui, T.S. Huang, and S. Mehrotra. Exploring video structure beyond the shots. In *International Conference on Multimedia Computing and Systems*, pages 237–240, 1998.

275. Y. Rui, T.S. Huang, and S. Mehrotra. Constructing table-of-content for videos. *ACM Multimedia Systems Journal, Special issue Multimedia Systems on video librairies*, 1999. To appear.

276. Yong Rui, Thomas S. Huang, and Sharad Mehrotra. Constructing table-of-content for videos. *Multimedia Syst.*, 7(5):359–368, 1999.

277. D. Sadlier, S. Marlow, N. O'Connor, and N. Murphy. Automatic tv advertisement detection from mpeg bitstream. *Journal of the Pattern Recognition Society*, 35(12):2719–2726, 2002.

278. P. Salembier, R. Qian, N. O'Connor, P. Correia, I. Sezan, and P. van Beek. Description schemes for video programs, users and devices. *Signal Process.: Image Commun.*, 16:211–234, 2000.

279. G. Salton and M.J. McGill. *Introduction to modern information retrieval*. McGraw Hill, New York, 1983.

280. J. Saunders. Real-time discrimination of broadcast speech/music. *ICASSP'96*, II:993–996, Atlanta, May 1996.

281. D.D. Saur, Y-P. Tan, S.R. Kulkarni, and P. Ramadge. Automatic analysis and annotation of basketball video. In *SPIE Conference on Storage and Retrieval for Still Image and Video Databases*, pages 176–187, 1997.

282. S. Sav, H. Lee, A.F. Smeaton, N. O'Connor, and N. Murphy. Using video objects and relevance feedback in video retrieval. In *SPIE Vol. 6015*, 2005.

283. N. Sawhney, D. Balcom, and I. Smith. Hypercafe: narrative and aesthetic properties of hypervideo. In *HYPERTEXT '96: Proceedings of the the seventh ACM conference on Hypertext*, pages 1–10, New York, NY, USA, 1996. ACM Press.

284. E. Scheirer and M. Slaney. Construction and evaluation of a robust multifeature speech/music discrimination. *ICASSP'97*, 4, Munich, Germany, 1997.

285. H. Schneiderman. *A statistical approach to 3D object detection applied to faces and cars*. PhD thesis, Robotics Institute, Carnegie Mellon University, 2000.

286. H. Schneiderman and T. Kanade. A statistical method for 3d object detection applied to faces and cars. *IEEE Conf. on Computer Vision and Pattern Recognition*, pages 746–751, 2000.

287. B. Schölkopf and A. Smola. *Learning with kernels.* MIT Press, Cambridge, MA, 2002.

288. G. Schwarz. Estimating the dimension of a model. *The Annals of Statistics*, 6(2):461–464, March 1978.

289. M. Seitz and C.R. Dyer. View-invariant analysis of cyclic motion. *International Journal of Computer Vision*, 25:1–25, 1997.

290. G. Shakhnarovich, J.W. Fisher, and T. Darrel. Face recognition from long-term observations. In *Proc. IEEE European Conference on Computer Vision*, 3:851–868, 2002.

291. F. Shipman, A. Girgensohn, and L. Wilcox. Hypervideo expression: experiences with hyper-hitchcock. In *HYPERTEXT '05: Proceedings of the sixteenth ACM conference on Hypertext and hypermedia*, pages 217–226, New York, NY, USA, 2005. ACM Press.

292. F.M. Shipman, A. Girgensohn, and L. Wilcox. Hyper-Hitchcock: Towards the easy authoring of interactive video. In *Human-Computer Interaction INTER-ACT '03: IFIP TC13 International Conference on Human-Computer Interaction.* IOS Press, Sep. 2003.

293. T. Sikora. The mpeg-4 video standard verification model. *IEEE Trans. Circuits Systems Video Technol.*, 7(1):19–31, 1997.

294. Q. Siskind, J.M. and Morris. A maximum likelihood approach to visual event classification. In *European Conference on Computer Vision*, pages 347–360, 1996.

295. J. Sivic, M. Everingham, and A. Zisserman. Person spotting: video shot retrieval for face sets. *International Conference on Image and Video Retrieval*, 2005.

296. A.F. Smeaton. Indexing, Browsing and Searching of Digital Video. *Annual Review of Information Science and Technology (ARIST)*, 38:371–407, 2004.

297. A.F. Smeaton, J. Gilvarry, G. Gormley, B. Tobin, S. Marlow, and M. Murphy. An evaluation of alternative techniques for automatic detection of shot boundaries in digital video. In *Irish Machine Vision and Image Processing Conference*, Dublin, Ireland, September 1999.

298. A.F. Smeaton, C. Gurrin, H. Lee, K. Mc Donald, N. Murphy, N. O'Connor, D. O'Sullivan, and B. Smyth. The Físchlár-News-Stories System: personalised access to an archive of TV news. In *RIAO2004*, 2004.

299. A.F. Smeaton, N. Murphy, N. O'Connor, S. Marlow, H. Lee, K. Mc Donald, P. Browne, and J. Ye. The físchlár digital video system: A digital library of broadcast TV programmes. In *ACM+IEEE Joint Conf. on Digital Libraries*, 2001.

300. Alan F. Smeaton, Paul Over, and Wessel Kraaij. TRECVid: evaluating the effectiveness of information retrieval tasks on digital video. In *MULTIMEDIA '04: Proceedings of the 12th annual ACM international conference on Multimedia*, pages 652–655, New York, NY, USA, 2004. ACM Press.

301. A. Smeulders, M. Worring, S. Santini, and A. Gupta. Content based image retrieval at the end of the early years. *IEEE Transaction on Pattern Analysis and Machine Intelligence*, 22(12):1349–1380, 2000.

302. A.W.M. Smeulders, M. Worring, S. Santini, A. Gupta, and R. Jain. Content-based image retrieval at the end of the early years. *IEEE Trans. Pattern Anal. Machine Intell.*, 22:1349–1380, December 2000.

303. J.M. Smith, D. Stotts, and S.-U. Kum. An orthogonal taxonomy for hyperlink anchor generation in video streams using OvalTine. In *HYPERTEXT '00: Proceedings of the eleventh ACM on Hypertext and hypermedia*, pages 11–18, New York, NY, USA, 2000. ACM Press.

304. M.A. Smith and T. Kanade. Video skimming and characterization through the combination of image and language understanding. In *IEEE International Conference on Computer Vision and Pattern Recognition*, pages 775–781, 1997.

305. M.A. Smith and T. Kanade. Video skimming and characterization through language and image understanding techniques. Technical report, Carnegie Mellon Univ., 1995.

306. S. Smoliar and H. Zhang. Content-based video indexing and retrieval. *IEEE Multimedia Magazine*, 1(2):62–72, 1994.

307. C.G.M. Snoek, M. Worring, J.M. Geusebroek, D.C. Koelma, and F.J. Seinstra. The MediaMill TRECVID 2004 semantic video search engine. In *TREC Video Retrieval Evaluation Online Proceedings*, 2004.

308. M. Sonka, V. Hlavac, and R. Boyle. *Image Processing, Analysis, and Machine Vision, Second Edition*. Brooks/Cole, 2002.

309. S. Srinivasan, D. Petkovic, and D. Ponceleon. Towards robust features for classifying audio in the CueVideo system. *ACM Multimedia'99*, 1999.

310. C. Stauffer and W. Grimson. Learning patterns of activity using real-time tracking. *IEEE Transactions on Pattern Recognition and Machine Intelligence*, 22:747–757, 2000.

311. A. Stefanidis, A. Partsinevelos, and A. Doucette. Summarizing video data-sets in the spatiotemporal domain. In *11th International Workshop on Database and Expert Systems Applications*, pages 906–912, 2000.

312. A. Steinmetz. Media and distance: A learning experience. *IEEE Multimedia*, pages 8–10, 2001.

313. Thomas Strohmer. Approximation of dual gabor frames, window decay, and wireless communications. *Appl.Comp.Harm.Anal.*, 11(2):243–262, 2001.

314. H. Sundaram and S. Chang. Video skims: Taxonomies and an optimal generation framework. In *IEEE International Conference on Image Processing (ICIP'02)*, Rochester, NY, 2002.

315. H. Sundaram and S.-F. Chang. Condensing computable scenes using visual complexity and film syntax analysis. In *Proceedings of the 2001 IEEE International Conference on Multimedia and Expo, ICME 2001*. IEEE Computer Society, Aug. 2001.

316. K.K. Sung and T. Poggio. Example-based learning for view-based human face detection. *IEEE Transactions on Pattern Analysis and Machine Intelligence*, 20(1):39–51, 1998.

317. R. Swan and J. Allan. Timemine: visualizing automatically constructed time-lines. In *SIGIR2000*, page 393, 2000.

318. H. Tao, H.S. Sawhney, and R. Kumar. Dynamic layer representation and its applications to tracking. In *Proc. IEEE Int'l Conf. on Comput. Vis. and Patt. Recog.*, pages 134–141, June 2000.

319. C.M Taskiran, A. Amir, D.B. Ponceleon, and E.J. Delp. Automated video summarization using speech transcripts. In *SPIE Storage and Retrieval for Media Databases*, pages 371–382, 2001.

320. D. Terzopoulos and R. Szeliski. Tracking with Kalman snakes. In *Active Vision*, pages 3–20. Cambridge, MA: MIT Press, 1992.

321. P.A. Tipler. *Physik*. Spektrum Akademischer Verlag, Heiderlberg, Berlin, 1994.

322. C. Toklu and S.P. Liou. Semi-automatic dynamic video object marker creation. In *Proc. of SPIE Storage and Retrieval for Image and Video Databases VII* Vol.*3656*, San Jose, Cal., USA, January 1999.

323. Y. Tonomura, A. Akutsu, K. Otsuji, and T. Sadakate. Videomap and videospaceicon: Tools for anatomizing video content. In *INTERCHI'93*, pages 131–141, 1993.

324. TREC video retrieval evaluation.
http://www-nlpir.nist.gov/projects/trecvid/.

325. TRECVid. TREC Video Retrieval Evaluation. website available at:
http://www-nlpir.nist.gov/projects/t01v/t01v.html (visited august 2005).

326. B.T. Truong and S. Venkatesh. Video abstraction: A systematic review and classification. *ACM Transactions on Multimedia Computing, Communications, and Applications*, May 2005.

327. S. Uchihashi, J. Foote, A. Girgensohn, and J. Boreczky. Video Manga: generating semantically meaningful video summaries. In *MULTIMEDIA '99: Proceedings of the seventh ACM international conference on Multimedia (Part 1)*, pages 383–392, New York, NY, USA, 1999. ACM Press.

328. H. Ueda, T. Miyatake, and S. Yoshizawa. An interactive natural motion picture dedicated multimedia authoring system. In *ACM SIGCHI 91*, pages 343–350, 1991.

329. P. Van Beek and J. Smith. *Introduction to MPEG-7*, chapter Navigation and Summarization. Wiley, 2002.

330. Rudolph van der Merwe, A. Doucet, N. Freitas, and E. Wan. The unscented particle filter. Technical Report CUED/F-INFENG/TR 380, Cambridge University Engineering Department, 2000.

331. Rudolph van der Merwe and Eric A. Wan. Sigma-point kalman filters for probabilistic inference in dynamic state-space models. In *Proc. of the Workshop on Advances in Machine Learning*, June 2003.

332. Rudolph van der Merwe and Eric A. Wan. Sigma-point kalman filters for integrated navigation. In *Proc. of the 60th Annual Meeting of The Institute of Navigation (ION)*, June 2004.

333. T. Veit, F. Cao, and P. Bouthemy. An a contrario decision framework for motion detection. In *IEEE Int. Conf on Computer Vision and Pattern Recognition (CVPR'2004)*, Washington, USA, June 2004.

334. J. Vermaak and A. Blake. Nonlinear filtering for speaker tracking in noisy and reverberant environments. In *Proc. IEEE Int'l Conf. Acoustic Speech Signal Processing*, pages V:3021–3024, 2001.

335. J. Vermaak, A. Blake, M. Gangnet, and P. Perez. Sequential Monte Carlo fusion of sound and vision for speaker tracking. In *Proc. IEEE Int'l Conf. on Computer Vision*, pages 741–746, 2001.

336. P. Viola and M. Jones. Robust real-time face detection. *International Journal of Computer Vision*, 57(2):137–154, 2004.

337. V. Wan and W.M. Campbell. Support vector machines for speaker verification and identification. *Proc. of the IEEE Signal Processing Society Workshop on Neural Networks*, 2:775–784, 2000.

338. H. Wang and P. Chu. Voice source localization for automatic camera pointing system in video conferencing. In *Proc. IEEE Int'l Conf. Acoustic Speech Signal Processing*, 1997.

339. B.M. Wildemuth, G. Marchionini, M. Yang, G. Geisler, T. Wilkens, A. Hughes, and R. Gruss. How fast is too fast?: evaluating fast forward surrogates for digital video. In *JCDL '03: Proceedings of the 3rd ACM/IEEE-CS joint conference on Digital libraries*, pages 221–230, Washington, DC, USA, 2003. IEEE Computer Society.

340. A.D. Wilson, A.F. Bobick, and J. Cassell. Temporal classification of natural gesture and application to video coding. In *IEEE Conference on Computer Vision and Pattern Recognition*, pages 948–954, 1997.

341. W. Wolf. Key frame selection by motion analysis. In *IEEE Int Conf Acoust, Speech, and Signal Proc*, 1996.

342. Patrick J. Wolfe, Simon J. Godsill, and Monika Dörfler. Multi-gabor dictionaries for audio time-frequency analysis. *Proceedings of WASPAA 2001*, 2001.

343. Y. Wu and T.S. Huang. Color tracking by transductive learning. In *Proc. IEEE Int'l Conf. on Comput. Vis. and Patt. Recog.*, pages I:133–138, 2000.

344. Y. Wu and T.S. Huang. View-independent recognition of hand postures. In *Proc. IEEE Int'l Conf. on Comput. Vis. and Patt. Recog.*, pages II:88–94, 2000.

345. J. Xi, X.-S. Hua, X.-R. Chen, L. Wenyin, and H.-J. Zhang. A video text detection and recognition system. In *IEEE International Conference on Multimedia and Expo (ICME)*, Waseda University, Tokyo, Japan, August 22-25 2001.

346. L. Xie and S.-F. Chang. Structure analysis of soccer video with hidden markov models. In *IEEE International Conference on on Acoustics, Speech, and Signal Processing*, 2002.

347. W. Xiong, C.J Lee, and M. Ip. *Storage and retrieval for image and video databases*, chapter Net comparison: a fast and effective method for classifying image sequences, pages 318–326. February 1995.

348. Z. Xiong, R. Radhakrishnan, and A. Divakaran. Generation of sports highlights using motion activity in combination with a common audio feature extraction framework. In *IEEE International Conference on Image Processing (ICIP'03)*, pages 5–8, Barcelona, Spain, 2003.

349. I. Yahiaoui. *Construction automatique de résumés videos*. PhD thesis, Telecom Paris - Institut Eurecom, 2003.

350. I. Yahiaoui, B. Merialdo, and B. Huet. Generating summaries of multi-episodes video. In *ICME 2001, International Conference on Multimedia and Expo*, Tokyo, Japan, August 2001.

351. I. Yahiaoui, B. Merialdo, and B. Huet. Comparison of multi-episode video summarization algorithms. *EURASIP Journal on applied signal processing Special issue on multimedia signal processing*, 2003(1):48–55, January 2003.

352. I. Yahiaoui, B. Mérialdo, and B. Huet. Optimal video summaries for simulated evaluation. In *CBMI 2001 European Workshop on Content-Based Multimedia Indexing*, Brescia, Italy, 2001.

353. Yahoo! video search. http://video.search.yahoo.com.

354. A. Yamada, M. Pickering, S. Jeannin, L. Cieplinski, J.R. Ohm, and M. Kim. Mpeg-7 visual part of experimentation model version 9.0, 2001.

355. C.Y Yang and J.C. Lin. Rwm-cut for color image quantization. *Computer and Graphics*, 20(4):577, 1996.

356. B.L. Yeo and B. Liu. Rapid scene analysis on compressed video. *IEEE Trans. on Circuits and Systems for Video Technology*, 6(5):533–544, 1995.

357. M. Yeung and B. Yeo. Segmentation of video by clustering and graph analysis. *Computer Vision and Image Understanding*, 71(1):94–109, July 1998.

358. M.M. Yeung and B.-L. Yeo. Video visualization for compact presentation and fast browsing. *IEEE Transactions on Circuits and Systems for Video Technology*, 7(5):771–785, Oct. 1997.

359. L. Ying, T. Zhang, and D. Tretter. An overview of video abstraction techniques. Technical report, Imaging Systems Laboratory HP Laboratories Palo Alto, January 2001. Technical Report: HPL-2001-191.

360. Ian T. Young, Lucas J. van Vliet, and Michael van Ginkel. Recursive gabor filtering. *IEEE Transactions on Signal Processing*, 50:2798–2805, 2002.

361. D. Yow, B-L. Yeo, M. Yeung, and B. Liu. Analysis and presentation of soccer highlights from digital video. In *2nd Asian Conference on Computer Vision*, 1995.

362. X.D. Yu, L. Wang, Q. Tian, and P. Xue. Multi-level video representation with application to keyframe extraction. In *International Conference on Multimedia Modeling (MMM'04)*, pages 117–121, Brisbane, Australia, 2004.

363. Z. Yu, Z. Hongjiang, and A.K. Jain. Automatic caption localization in compressed video. *IEEE Trans. On PAMI*, 22(4):385–392, April 2000.

364. R. Zabih, J. Miller, and K. Mai. A feature-based algorithm for detecting and classifying scene breaks. In *3rd International Multimedia Conference and Exhibition, Multimedia Systems*, pages 189–200, 1995.

365. H. Zettl. *Sight, Sound, Motion: Applied Media Aesthetics*. 3rd Edition, Wadsworth Pub Company, 1999.

366. D. Zhang, R.K. Rajendran, and S.-F. Chang. General and domain-specific techniques for detecting and recognizing superimposed text in video. In *International Conference on Image Processing*, pages 22–25, Rochester, New York, USA, September 2002.

367. H.J. Zhang, S.Y. Tan, S.W. Smoliar, and G. Yihong. Automatic parsing and indexing of news video. *Multimedia Syst.*, 2(6):256–266, 1995.

368. H.S. Zhang and H.J. Wu. Content-based video browsing tools. *In Multimedia Computing and Networking*, pages 389–398, 1995.

369. H.J. Zhang, C.Y. Low, S.W. Smoliar, and J.H. Wu. Video parsing, retrieval and browsing: An integrated and content-based solution. *ACM Multimedia*, pages 15–24, 1995.

370. T. Zhang and C.-C. Kuo. Audio content analysis for on-line audiovisual data segmentation. *IEEE Transactions on Speech and Audio Processing*, 9(4):441–457, 2001.

371. Z. Zhang, S. Furui, and K. Ohtsuki. On-line incremental speaker adaptation with automatic speaker change detection. *ICASSP'00*, 2000.

372. W. Zhao, R. Chellappa, P.J. Phillips, and A. Rosenfeld. Face recognition: A literature survey. *ACM Computing Surveys*, 35(4):399–458, 2004.

373. D. Zhong and S.-F. Chang. Structure analysis of sports video using domain models. In *IEEE International Conference on Multimedia & Expo*, 2001.

374. Y. Zhong, K. Karu, and A.K.Jain. Locating text in complex color images. *Pattern Recognition*, 28(10):1523–1536, Oct 1995.

375. Y. Zhong, H. Zhang, and A. Jain. Automatic caption localization in compressed video. *IEEE Transactions on Pattern Analysis and Machine Intelligence*, 22(4):385–392,, 2000.

376. Y. Zhuang, Y. Rui, T. Huang, and S. Mehrotra. Adaptive key frame extraction using unsupervised clustering. In *International Conference on Image Processing (ICIP'98)*, pages 866–870, Chicago, Illinois, 1998.

377. Y. Zhuang, Y. Rui, T.S. Huang, and S. Mehrotra. Adaptive key frame extraction using unsupervised clustering. In *Proc. of IEEE conf. on Image Processing*, pages 866–870, Chicago, IL, 1998.

378. Z. Zivkovic and B. Krose. An EM-like algorithm for color-histogram-based object tracking. In *IEEE Int. Conf on Computer Vision and Pattern Recognition (CVPR'2004)*, pages 798–803, Washington, USA, June 2004.

379. E. Zwicker and H. Fastl. *Psychoacoustics. Facts and Models.* Springer Verlag, second updated edition, 1999.

Index

MPEG-7, 134

ABC, 199
advertisement detection, 194
archives, 3
audio, 8
audio analysis, 166
audio feature, 168
audio objects, 110, 124
authoring, 20
automatic, 9
Automatic Speech Recognition, 200
automatic video structuring, 192

Bayesian estimation, 49
Bayesian Information Criterion, 62
Bayesian sequential estimation, 51
BBC, 189
browsing, 16, 45

Camera shot, 7
CCTV, 189
classification, 13
closed-caption, 195
cluster, 8
clustering, 22
clutter
    removal, *see* face segmentation
CNN, 199
Color histogram, 50
component, 9
Computational media aesthetics, 162
content, 3, 4, *see* retrieval
content management, 159, 160, 184

content selection, 40
cursor, 20
cut, 9

detail-on-demand, 4, 207, 209–212
detection, *see* face
    eye, 92, 93, 95
    mouth, 92, 93, 95
dissolves, 7
distance, 90
    robust, 92, 100
dual window, 111
dynamic programming, 75

e-learning, 159, 160, 184
edit, 20
EM algorithm, 61
event, 8, 14
Event detection, 140
event detection, 154
Event modeling, 133, 140
event modeling, 154
expression, *see* face
external, 17, 20
External-links, 17

face
    detection, 89
    expression, 90, 95, 100
    features, 91–93, 95, 99
    pose, 90–92, 95, 96
    recognition, 89, 101–105
    registration, 91, 92, 95, 99
    segmentation, 91, 92, 96–98

fades, 7
failure detection, 50
features, *see* face
films
    feature length, 89, 101–105
Físchlár-News, system, 194
Físchlár-TRECVid, system, 200

Gabor transformation, 108, 111
Gaussian window, 111
Gaussian window, 110, 112
Google, 3
Gradual transitions, 7

Heisenberg uncertainty principle, 112
Heisenberg uncertainty principle, 110,
    115
Hidden Markov Model, 139
hierarchical, 209, 210
high-level components, 9
highlight, 14
HMM
    contour tracking, 70
    transition probabilities, 72
    Viterbi algorithm, 72
home network, 40
human operators, 4
hyperfilm, 6
hyperlinks, 46
hypermedia, 6
HyperVideo, 24
hypervideo, 6, 25, 208–214, 217, 218,
    224

illumination, 90
    normalization, 91, 92, 98
image
    signature, 89, 91, 92, 98
imaged sound, 111
imaged sound, 119
indexing and retrieval, 133
information fusion, 142
instructional video, 164–168, 174, 184
integrated, 21
integration, 17
interactive, 25, 207, 208, 211, 212, 214,
    219, 224
interactive multimedia, 134
Interactive Video, 150

Interactive video, 5
interactive video, 4, 18–20
Interactive video , 5
Interactive video database, 6
interactive video database, 5
Interactive video presentation, 5
interactive video presentation, 5, 18
interface, 19, 20, 25, 207, 209–213, 216,
    219, 220, 224
internal, 17, 20
Internal-links, 17
intra-shot appearance changes, 58
iPods, 3
iTunes, 3
iVideoToGo, 20

joint probabilistic matching, 73

Kalman filtering, 53
Key-frame, 8, 10

latent semantic index, 221, 223
lattice constants, 111
link-opportunities, 18
links, 17
low-level components, 9

manual video annotation, 190
Markov random field, 49
matching, 45
    robust, *see* distance, robust
Maximum Recollection Principle, 27, 31
mean-shift, 51
Mixture of Gaussians, 57
mobileTV, 3
motion, 48
Movideo, 21
MPEG-1, 4, 5, 194
MPEG-2, 4
MPEG-4, 4
MPEG-7, 4, 150
MPEG-7 eXperimentation Model, 200
multi-episode video summarization, 30
multi-video summarization, 36
multimedia summaries, 40

narrative, 8
National Institute of Standards and
    Technology, 198

navigate, 20
navigation, 40, 45
navigational, 20
news story segmentation, 194
non-linear, 4, 5, 20
nonlinear dynamic systems, 77

object, 8, 11, 21
object detection, 48
object tracking, 67, 69, 71, 73, 75, 77,
    79, 81, 83, 85, 87
occlusion, 90, 92, 100
Online learning, 80
ordering
  rank score, 101

particle filtering, 53
personalization, 153
playback, 18
pose, *see* face
Principal Components Analysis, 58
probabilistic graph model, 138
probabilistic inference, 147

query-by-example, 16
Query-by-keywords, 16

raw interactive video, 5
re-structuring, 6, 9
recognition, 13, *see* face
registration, *see* face
resolution zooming, 115
resolution zooming, 110
retrieval, 208, 211
  content-based, 89, 101–105
RTE, 194

scene, 8
searching, 16, 45
selection mask, 120
semantic gap, 133, 160
Shot, 9
shot, 21
shot boundary detection, 194
SMIL, 24
space-state framework, 50

spectrogram, 107, 111
story-board, 29
structure, 6
Summaries, 27
summarization, 214
summary, 8
Support Vector Machines, 92, 93, 95
surveillance videos, 48
Synchronization, 18
synchronization, 143
Synchronized Multimedia Integration
    Language, 19
synthetic
    data augmentation, 93

table-of-contents, 6
template, 110, 124
Text, 15
text, 8
Text REtrieval Conference, 198
the move, 3
time-frequency resolution, 110, 115, 119
tracking, 13, 45
TRECVID, 219
TRECVid, video benchmarking, 198

Unscented Kalman Filter, 76

variational approaches, 49
VCR, 4
VeonStudioTM, 24
video OCR, 200
video ontology, 190
video summary, 27, 152
video-on-demand, 3
video-skims, 28
VideoClic, 20, 23
VideoPrep, 20
Visual Audio, 107

wipes, 7
World Wide Web Consortium, 19

Yahoo, 3

zone, 8, 11

Signals and Communication Technology

(continued from page ii)

**Chaos-Based Digital
Communication Systems**
Operating Principles, Analysis Methods,
and Performance Evalutation
F.C.M. Lau and C.K. Tse   ISBN 3-540-00602-8

Adaptive Signal Processing
Application to Real-World Problems
J. Benesty and Y. Huang (Eds.)
ISBN 3-540-00051-8

**Multimedia Information Retrieval
and Management**
Technological Fundamentals and Applications
D. Feng, W.C. Siu, and H.J. Zhang (Eds.)
ISBN 3-540-00244-8

Structured Cable Systems
A.B. Semenov, S.K. Strizhakov,
and I.R. Suncheley   ISBN 3-540-43000-8

UMTS
The Physical Layer of the Universal Mobile
Telecommunications System
A. Springer and R. Weigel
ISBN 3-540-42162-9

Advanced Theory of Signal Detection
Weak Signal Detection in
Generalized Obeservations
I. Song, J. Bae, and S.Y. Kim
ISBN 3-540-43064-4

Wireless Internet Access over GSM and UMTS
M. Taferner and E. Bonek
ISBN 3-540-42551-9